AF544959

HOW TO BUILD YOUR OWN WORKING ROBOT PET

Dedication

To the Creator, who in turn
taught men to be creative

No. 1141
$10.95

HOW TO BUILD YOUR OWN WORKING ROBOT PET

BY FRANK DaCOSTA

FIRST EDITION

FIRST PRINTING—MAY 1979

Printed in the United States of America

On the cover: NASA thermal image by Hugh M. Kieffer.

Library of Congress Cataloging in Publication Data

DaCosta, Frank.
How to build your own working robot pet.

1. Automata. I. Title.
TJ211.D32 629.8'92 79-14243
ISBN 0-8306-9796-9
ISBN 0-8306-1141-X pbk.

Preface

Of all theoretical themes of the Twentieth Century, few are subject to as much recurring popularity as the subject of robots. Every few years, this intriguing concept experiences a resurgence of interest, due to current science fiction books, magazines, movies and television shows about it. And usually, the popular conception of a "robot" is a shiny, metallic humanoid, who utters a monotonic "Yes, Master" and proceeds to serve dinner. The robot is the modern symbol of the next Technological Revolution, the promise of a bright future.

So why, of all things, build a robot *dog*?

The fact of the matter is that the popular concept of the robot is a good deal different from the reality. Scores of experimenters today are developing prototype robots in their workshops, but their creations are far from the video versions. The robots of science fiction threaten to rival man in intelligence and physical ability; the actual prototypes have a hard time finding their way through doors.

The functions of a human house-servant are really quite complex, and a robot counterpart to replace him would be impractical for a dozen reasons. The functions of a pet, however, are few and simple, and a robot pet is not at all inconceivable. A robot butler would be much more costly than a human butler, and would do the job about 40 percent as well. The dog, though, is both a good robot *and* a good pet. It is for this reason that I decided to trade in the "golden-armored" robot for a little grey-furred machine with an electronic "bark."

The robot pet "taking a nap."

I would like to express gratitude to a number of individuals and agencies who helped bring both the robot pet and this volume to completion. In the "individuals" category, many thinks go to Richard Ruggles, whose tools and mechanical expertise were indispensable in the early stages of construction, and to my wife Cheryl, who typed this manuscript and learned patience thereby. In the "agencies" category, I thank the various members of the United States Robotics Society and the contributors to *Robot Builder* newsletter for pioneering work and encouragement; and I offer a special word of appreciation to the Intel Corporation for their high standard of excellence, which served as a goal for this project.

Frank DaCosta

Contents

1 **Birth of a Robot Pet** 9
Definition—Using This Volume.

2 **Overview: The Task at Hand** 16
Motorframe—Mainframe—Programming—Battery Charger.

3 **Building the Body** 23
Design Considerations—Choice of Motors—Mounting the Motors—Wheels—Final Test.

4 **Interfacing the Body** 36
Selection of Relays—Mounting the Relays—Servo Potentiometer—Connectors.

5 **The Power System** 48
Voltage Requirements—Current Requirements—The Voltage Regulator—Checking the Power System.

6 **The Charging System** 57
Feeding-Time Approaches—Charging System.

7 **The RCU-85: A Brain for the Body** 65
Processor Interaction—The RCU-85— Addressing Brain Board Assembly.

8 **Servodrive** 84
Servodrive Description—Circuit Wiring.

9 **Soniscan** 93
Soniscan Design—Transducers—Circuit Description—Alignment Procedure—Soniscan in Operation.

10 **Excom** 108
Fredian Grammar—Circuit Construction and Testing.

11 **Audigen**..........121
Design Considerations—Circuit Description—Circuit Assembly—Circuit Testing—Summary.

12 **And Other Circuits**..........130
Battery Monitor—Event Timer—Contact Switches—Various Fine Touches—Moving On.

13 **Manual Programmer**..........138
Circuit Description—Alternate Programmer—Construction—Final Testing.

14 **Tape Interface**..........147
Design Considerations—Circuit Description—How It Operates—Storage Function—Retrieval Function—Display Stage—Construction and Testing—Summary.

15 **Programming Techniques**..........168
The Fear of Software—The Processor and Instructions—The Instruction Set—Using the Instruction Set—Further Instructions.

16 **Mech 1: The Software Package**..........185
Initialization—Interrupt Addresses—Battery Monitor Routine—Excom Routine—ARASEM Routine—Callable Routines—General Subroutines—Final Programming Notes.

17 **Where From Here?**..........209
Toward Better Locomotion—Toward Improved Memory—Toward Enhanced Software—A Robot Pet In Every Home?

Appendices

A. Possible Motor Sources for the Robot Pet..........217
B. Powers of Two..........218
C. Hexadecimal-Decimal Integer Conversion..........219

Index..........231

Chapter 1
Birth of a Robot Pet

It started innocently enough, during the good humor of a summer evening. My wife and I were pondering the fact that our apartment complex would not permit us to own a pet dog. Half in jest (little did I know!), I boldly countered, "Well, if we cannot have a *real* dog, we'll just *build* one—a robot dog!" You can guess the flurry of wit that followed— "Electricity is cheaper than dog food!"—and yet, there was anticipation. Could such a project be realized?

This volume presents the entire design and construction of a robot pet. Anyone who tackles this project, though, will gain far more than a housepet. For the robot pet is a sort of tutorial experience, a tribute to three related fields of science. The first is the field of mechanics, the basis for the body of the pet. The second is the field of electronics, especially as manifested in the advent of the Microcomputer Revolution, the source for the brain of the pet. The third is the infant study known as robotics, the foundation for the design philosophy of the pet.

DEFINITION

What is a robot? A steady diet of science-fiction may leave you with some misconceptions. Simply speaking, a robot is an *electromechanical simulation of animal life.* That is, a robot is a machine which exhibits many or most of the characteristics evident in animals and in man. This definition has a direct bearing on the design of the robot pet and similar automata.

Living things demonstrate a host of hard-to-simulate characteristics. Just to give an idea, consider the following:

- Animals are aware of their environment through a variety of senses.
- They move around in that environment.
- They manipulate it with their limbs.
- They communicate with others in it.
- They sustain themselves in it by consumption of food.
- They coordinate all these activities according to some system of decision-making and "prioritizing."

A machine becomes more worthy of the term "robot" as it manages to genuinely integrate more of these living characteristics into its design. Figure 1-1 attempts to picture these traits as they relate to one another.

The pet robot, by the very nature and purpose of its design, is built to exhibit each of these characteristics to one or another degree. Thus, the completed pet will be quite "lifelike," because the major observable qualities of life are demonstrated. In addition, the high degree of integration involved in the robot pet system ensures that all of the characteristics interact, in much the same way that they do in true life. That is, the pet robot is not a "puppet," manipulated to *look like* it is making logical, lifelike decisions; it is truly playing out lifelike processes, though on a much lower level.

The robot pet, then, like a living thing, is aware of its environment. We are familiar with the five senses in Man and animals, the primary one being vision. The pet robot uses a heightened sense of hearing, coupled with the ability to emit ultrasonic pulses, as its primary sense. It is termed *Soniscan* in this volume, but it is in reality merely an electronic version of the same sort of thing that bats and dolphins have been doing for many years—navigation by echo-detection. The pet robot cannot "see" in the conventional sense, but Soniscan is enough to prevent collision with obstacles.

The pet robot can move around in its environment. We're used to seeing the family dog jump down from the couch and run to the front door, using those embodiments of agility, the legs. The pet robot has no legs, but makes use of a strong front-wheel drive mechanism to get around. The mechanism allows forward or reverse motion and some 16 different turning-angles, all under the full control of the pet's "brain." It may not replace legs, but it's maneuverable enough to get the pet in and out of tight spots. The robot pet can nominally manipulate its environment as well, since the drive mechanism is strong enough to push small obstacles in its path.

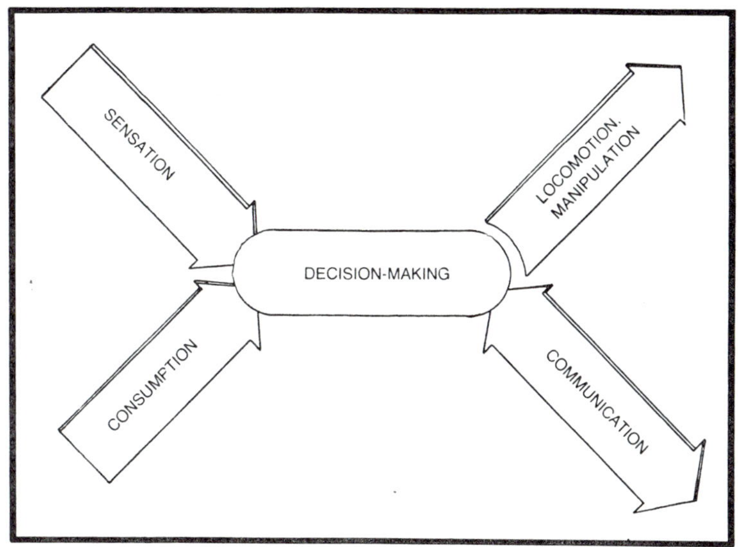

Fig. 1-1. The "life characteristics" which a robot must demonstrate.

The robot pet can communicate with others in its environment. We expect our dogs to bark, our cats to meow. And through careful observation, we learn that this is an "I want to go outside" bark or an "I'm hungry" meow. The pet robot, by means of its *Audigen* (for audio generation) circuit, has a set of synthesized sounds with which to indicate its "moods." And since the builder/programmer determines what triggers these sounds, they can carry much more information than the typical dog's bark, approaching a set of "audible diagnostics."

In turn, we send our pets to obedience school to "teach them to listen." We expect our dogs to stop barking when we tell them to do so. The robot pet can hear *and* understand vocal commands given it by its owners, by means of a circuit called *Excom* (for EXternal COMmand). It requires that the owner learn to speak the pet's language a bit, but it is at least a spoken language—the owner doesn't require a remote control box. And once again, since the owner is the programmer, he can attach definitions to the new language at will, making the Excom an effective means of communications.

The robot pet requires sustenance in its environment to continue on. People and their pets recharge their energy cells by the complex chemical processes involved in eating. The robot pet operates directly from electrical energy stored in its lead-acid batteries. "Mealtime" for it is a trip to the charger to restore its levels for renewed activity. The majority of the pet's weight belongs to the *power system*, but if it's a choice between mobility with extra weight versus a line-cord umbilical with less weight, I advocate the more "lifelike" choice—mobility.

Finally, the pet robot orders all of its activities according to a centralized, prioritized goal system. In life, we have the example of the brain, a highly-complex data center which engages itself in four main functions:

- Remembering
- Perceiving
- Deciding
- Controlling

There are many electronic circuits which can be used to perform one or a few of these tasks. But there is one sort of circuit which can handle them all, in a manner analogous to a living brain—and that is the microprocessor. The robot pet makes use of the Intel 8085A, one of the most powerful 8-bit microprocessors available today. A basic microcomputer is built around this processor to form the Robot Control Unit-85 (RCU-85). The RCU-85 can remember by using its random access memory; it can perceive with external TTL circuits interfaced to its input ports; it can decide by means of its powerful instruction set and versatile programming philosophy; and it can control with external circuits accessed by its output ports. In complexity, it is far from a living brain; in operation, however, it is quite similar.

So, then, the pet robot is a machine designed to simulate life in its primary aspects. Consequently, the person attempting its construction should be aware of the task ahead. Life is complex; pseudo-life must reflect this complexity somewhat. The pet robot is not the sort of thing one builds as a toy for his son. It is a difficult project, if only from the standpoint of time involved. A few words should be addressed, then, to the proper use of this volume.

USING THIS VOLUME

First of all, it is entirely possible that some will read this book and incorporate only a key circuit or two into a homebrew robot completely unlike the robot pet. This is perfectly legitimate, for not everyone wants a "robot pet," and some of the pet robot design is a matter of taste rather than necessity. With this in mind, I've tried to treat the separate circuits in semi-isolation, in enough detail to allow modified use in other robot systems. In most of these chapters, I've spent a reasonably large amount of time describing the "design cycle"—that is, the rationale leading up to the choices actually incorporated in the prototype. In this way, the options are laid open to the reader, who may then choose alternate approaches in his personal design. To this degree, this volume may be handled as a sort of "Robot Design Sourcebook."

A particular disadvantage, inherent in the robot pet project is related to its chief *advantage*—its centralization. Everything is coordinated by the ***brain board***, and specifically, by the RCU-85 microcomputer. As a result, many of the circuits, particularly Soniscan, are not fully testable until the project is largely complete. In practical terms, this means that the little "landmarks" and rewards along the way are few, and it is easy to become discouraged and drop the project altogether. However, many construction projects are similar to this one, particularly in the related field of personal computing. Hopefully, the knowledge of the eventual outcome of the project will be enough to keep the serious experimenter going.

What about tools? Remembering that the field of robotics is a union of the mechanical and the electronic, the reader should have access to tools of both kinds. In construction of the body of the robot pet, a considerable amount of machine shop-type work will be involved. Thus, tools such as hand drills (or a drill press), a table saw, and a bench press for forming aluminum are valuable. From the electronics aspect, it will be all but impossible to get through the project without a decent volt-ohmmeter of some sort, perhaps a +5-volt bench power supply, and a good oscilloscope to verify waveforms. In some cases, an all-purpose frequency generator could be helpful.

The circuit philosophy for the brain board of the robot pet is basic TTL. Aside from the high-level microprocessor-related chips, which themselves are TTL-compatible, most other ICs have been drawn from the standard 7400 series of digital circuits. Some linear circuits, especially the 558 dual operational amplifier (op amp), the 555 timer, the 311 voltage comparator, and the 567 tone decoder, also play important parts throughout the system. In order to minimize current requirements (unless you prefer a "warm" pet), the digital ICs have been chosen from the low-power Schottky (LS) branch of the 7400 series, though the circuit designations in the diagrams will appear in the basic number format, without the LS interfix, for simplicity. If you wish to make substitutions for one or another reason, be careful to read the related discussion to determine that the proposed substitute IC will meet the requirements of timing or level. Table 1-1 has been included to aid in the proper wiring of circuits.

A word on socket wiring is necessary at this point. The robot pet makes use of some of the most complex electronic circuitry currently available to the average experimenter. Fortunately enough, the circuit designers (particularly in the case of the 8085A) have gone out of their way to make their circuits easy to hook up and use. The level of sophistication, though, demands some care in handling and wiring. A case in point is that wire-wrapping is universally used in the construction of the brain board; solderable jumper leads would be impractical and unreliable. Both manual and battery-powered wire-wrap tools are readily available at computer shops and through mail-order houses advertised in electronics

Table 1-1. Table Of Power Pins For Integrated Circuits

CHIP#	+5 PIN	GND PIN
7400	14	7
7404	14	7
7410	14	7
7416	14	7
7420	14	7
7425	14	7
7430	14	7
7438	14	7
7442	16	8
7447	16	8
7473	4	11
7474	14	7
7475	5	12
7485	16	8
7490	5	10
7493	5	10
74123	16	8
74125	14	7
74139	16	8
74151	16	8
74157	16	8
74164	14	7
74175	16	8
311	8	4
386	6	4
555	8	1
558	8	4
567	4	7
8085A	40	20
8155	40	20

magazines. Learning to wire-wrap well takes a little effort, but the resultant skill is indispensable for circuit prototyping of any real complexity.

A final note may be helpful concerning parts placement. Because low-power signals are used in most cases on the brain board, most resistors may be chosen from the eighth-watt or quarter-watt variety; TTL voltage permit smaller capacitors and diodes, too. This being the case, it has been found helpful to bend and clip the leads of these parts, and insert them into standard 16-pin wire-wrap IC sockets for wiring. In this way, wire-wrapping can be used uniformly without needing to solder these other parts onto the board. In the brain board prototype, as a matter of fact, there are only two sorts of sockets: 40-pin wire-wrap sockets for the processor-related chips and 16-pin sockets for all else. Any 14-pin ICs merely leave a couple of socket-pins unused. Any 8-pin ICs, such as the 555 timer, use only half of the socket, leaving room for four assorted parts to be used with the timer. This technique makes for an organized layout and less confusion in the long run.

With these pertinent points taken care of, good luck on the robot pet project. With some perserverance and vision, it will not be long before your home echoes with the sound of exhuberant—albeit *electronic*—barking.

Chapter 2
Overview: The Task at Hand

In every project of magnitude, it is wise to step back, as it were, and take a preliminary overview of the task at hand. The robot pet is no quick endeavor; it involves a half-dozen interrelated sub-systems which must be understood in terms of their relationship to the whole. Hopefully, this chapter will put some "handles" on the project.

First of all, let's consider the robot proper. Figure 2-1 provides a side view without its usual furry "skin" covering. Though the reader's own version may vary somewhat from the prototype as pictured, many of the same features will be in evidence.

The first major task is the construction of the body, a tricycle-type movable cart. Now, it probably would have been possible to find a ready-built motorized platform for the prototype, but in all likelihood it would have been too bulky for our purposes. One of the keynotes in the construction of the robot is *goal-oriented design*—that is, priorities were set with respect to its size, weight, and general appearance, well before actual construction began. And so, not just any mobile cart would suffice; we wanted a pet, not a golf cart. The pet's body was not accidental, but the outgrowth of a deliberate attempt to move a drawing board design into reality.

This being the case, note that the resultant chassis in Fig. 2-2 is totally "custom." It consists of the *mainframe,* which supports the pet's batteries and brain board, and the *motorframe*, which supports the pet's drive and steering sub-assemblies. Eighth-inch-thick aluminum was cut, formed, and drilled using a standard tablesaw, a small hydraulic press, and a drill press. Aluminum is a good material

Fig. 2-1. The pet robot, minus its shaggy outer skin.

to work with, by the way, because it might be considered the "rich man's wood." It is durable and strong, but as workable as wood. The reader may wish to use steel, but be prepared for some sweat and frustration.

MOTORFRAME

The motorframe took some thought and experimentation, but now the results can be passed on to you in hope of saving some time and effort. The process by which the various motors and gearboxes were selected will be treated in a later chapter. But it is enough to say that the choice of motors cannot be haphazard. If you choose too large a motor, your power system suffers under the increased load; if you choose too small a motor, your pet may not be able to trudge through shag carpeting. Chapter 3 will provide some of the relevant gear-ratios and speeds as a reference guide within which to experiment.

Mounted above the motorframe is the "head" of the pet. The head holds the "sense organs" of the pet: notably, the ultrasonic transducers for its Soniscan system, and the microphone for its Excom system. (For this reason, the major angle-piece beneath the head is sometimes referred to as the "Sensorframe.") It is sensible to mount the transducers in this high location, so as to detect low-hanging obstacles, such as chair seats, which a low-mounted transducer might miss. The robot pet, at this stage, begins to take on an appearance analogous to a dog: narrow-based and tall. This may seem too unstable or top-heavy, but it has proven reliable. Plus, if the robot pet were low and broad, like a turtle or crab, it would have difficulty maneuvering in the tight corners and hallways of a small apartment. Again: goal-oriented design.

MAINFRAME

Within the confines of the mainframe are the power system and the brain board. The power system provides the necessary voltage for the brain board circuitry and for the motors. A connector on the undersurface of the pet's chassis, between the rear wheels, provides access to the batteries, for recharging. The brain board plugs into two connectors mounted above the batteries, and is thereby easily removable for upgrading or debugging. Another connector below the chassis accesses the brain board for programming and checking of "vital functions."

For all intents and purposes, the brain board *is* the robot pet. Theoretically, it could be plugged into any one of several properly-wired robot bodies and operate fully according to design. The major goal herein was *centralization*, the interrelation of all input, output, and decision-making functions. Thus, the board contains the RCU-85, which is a small control-oriented microcomputer, for program execution, as well as circuitry for the Excom and Soniscan senses, the Audigen "bark" output, and the Servodrive circuitry for control of the motorframe.

I have allowed a fair amount of space on the prototype brain board for expansion and improvement. As a matter of fact, the RCU-85 as I present it in this book is capable of a 50 percent memory

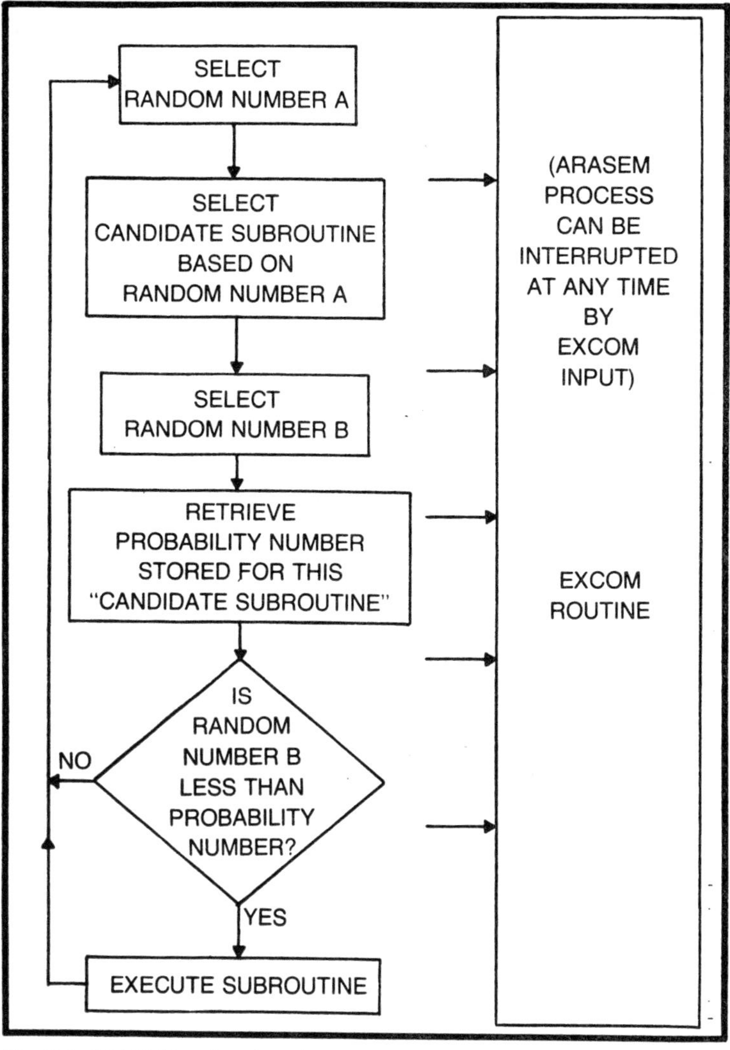

Fig. 2-2. ARASEM in flowchart form, with Excom as an interrupting program.

and Input/Output (I/O) expansion merely by the addition of one 40-pin IC. After all, anyone tackling something as unusual as a robotics project is by nature an experimenter. The robot pet as a "concept" may be a very high-level achievement, but certainly the prototype is only one step towards a "final generation" robot pet. And so, the more room for trial-and-error, the better, within the predetermined goals of size.

For that matter, because the entire brain board is a sort of plug-in module, the reader may even toy with other types of central processor. Before he does, though, let him heed the warnings noted in the chapter on the brain board. There is much to weigh before making such a sizable departure from the prototype design.

That is the robot pet, then, from a construction standpoint. Now, what can we expect from the machine? What will it be able to do once it is at least nominally completed?

PROGRAMMING

Those of you already familiar with microcomputer technology will understand when I say that the robot pet, in short, will *do what it is programmed to do*. There is the tendency to suppose, merely because of the speed and complexity of the microprocessor, that somehow the robot pet will have a "mind of its own." We may romanticize, we may personify, but the truth is, the pet robot is only a machine, capable only of operating within limited, preset parameters. It is finite, and therefore calculable: it may behave unexpectedly, but *not* "unpredictably." The burden is on the owner-programmer to tailor-make the personality of the pet.

Be that as it may, there is nothing less interesting than a puppet-on-a-string. There may be a sense of achievement in building a rolling machine that can be steered by remote control, but such a product just cannot be a pet. Even a microprocessor-based control system which is merely programmed to obey specific commands is not unique. What is needed, along the analogy of living things, is a *choice of goals*. That is, the pet robot must be obedient when commanded by the owner, but also capable of some limited pseudorandom activity of its own. The software package for the robot pet, discussed in Chapter 16, provides precisely this sort of choice.

When in full operating condition, the robot pet is normally in the *random mode*, under the control of a software monitor termed ARASEM, for ARtificially RAndom SElf Motivation. The ARASEM program selects from a number of user-programmed subroutines on a random basis, a la "roulette wheel." A probability table is maintained with a list of probability percentages to limit the change of execution of specific subroutines. In this manner, the owner can reinforce the chances of certain behavior and discourage other behavior—effectively shaping the personality of the pet.

While in the random mode, the robot may do some exploring, rolling about, always being careful to avoid obstacles, responding to changes in environment with "barks" or cautious approach. Or it may choose simply to sit for a few minutes, or to "temperamentally" growl at anything that moves. The completion of any particular goal returns the ARASEM routine back to the start for the selection of a new action.

The random mode may be interrupted by any command from the human operator, placing the robot pet into the Excom mode. The robot pet is programmed to recognize a limited vocabulary of vocally-produced command words in a special language described in Chapter 10. Upon the completion of any commanded action, the system reverts to the random mode, picking up where it left off before the interruption. This sort of automatic dual-mode routine allows the robot pet to operate with a minimum of human intervention, again reinforcing its similarity to a living pet. The necessary software techniques to support this process are described fully in Chapter 16, but Fig. 2-2 illustrates the concept.

Programming the robot pet is a straightforward task, once the software is written. To facilitate the programming, two special programmers are described in this book. The first is a simple manual programmer, which allows step-by-step loading of programs. Chapter 13 gives the necessary details for a few possible versions of the manual programmer, and you can choose one to fit your personal taste. The second programming tool is a *tape interface unit*, so that, once a program has been manually loaded, it may be stored on a standard cassette tape for instant restoration during the recharge cycle. This device is a "must," since the job of manual programming can be tedious.

BATTERY CHARGER

The pet's battery charger completes the project. It is designed to run for an average of twelve hours on one charge. When it senses its power supply running low, it begins a "hungry" sequence, in which it attempts to notify its owner of its need for a charge. After a given period of time, it voluntarily shuts down, lest it begin to act erratically due to low battery-voltages. It is the owner's task to "feed" the pet by means of the charger, which does the job automatically in a few hours, and to replenish the pet's memory with the cassette interface if the voltage-drop has brought about a loss of programming, which only takes a minute or two. (It may be said, and I will readily agree, that the analogy to life breaks down if the memory has to be periodically restored, and if the robot cannot seek out and use its own charger. And yet, even living pets are not so independent of their masters that they do not need the providing of food. The individual reader, though, may wish to experiment with some system of self-recharging; some pertinent observations are noted in Chapter 6.)

By far, the most formidable portion of the robot pet project is the physical construction of the chassis and drive mechanisms. After all, electronic schematics can be specified to the letter, but some ingenuity and experimentation are necessary when tailor-making part sizes and pulleys and selecting gear-motors. I have attempted to describe the delimiting factors concerning motors and such, plus the actual dimensions of parts that I used, in the hopes that any approximations or substitutions need not be random or directionless.

As a matter of fact, now that you've completed an overview of the task at hand, there is only one thing left to do: Read on into Chapter 3 and begin to construct the chassis.

Chapter 3
Building The Body

The only real factor of difference between a robot and a standard computer is their mode of expression. Computers manipulate data and express that activity through a cathode ray tube (CRT) or teletype. Robots likewise manipulate data and express that activity through a mobile extension. In short, a computer can "print out" its desires, but a robot can carry out those desires—by motion.

Most common animals make use of legs for locomotion, so in the early design stages of the pet robot, I did consider some form of legs (because of the analogy to life), and even made some good progress on paper. The primary stumbling block, alas, was steering. How does a dog, for instance, turn left or right? Watch one sometime; turning is not at all a simple process. There's a lengthening or shortening of the gait, and some lateral bending of the body. Noting this, I rejected legs as impractical for the time. (It has been done, of course, but by roboticists with far more time, intelligence, and resources than I can claim.) Regarding tank-treads and snake-coils as a bit incredible (though these, too, have been used successfully), I returned to the seemingly inescapable, age-old wheel.

The pet robot began, then, simply as a wheeled cart, as pictured in Fig. 3-1. But a number of considerations came into play as the concept grew on the drawing board. Let's see if we can handle them one by one.

DESIGN CONSIDERATIONS

First was the overall size. I wanted a pet-sized pet, one that would coexist comfortably with my wife and me. Our apartment is

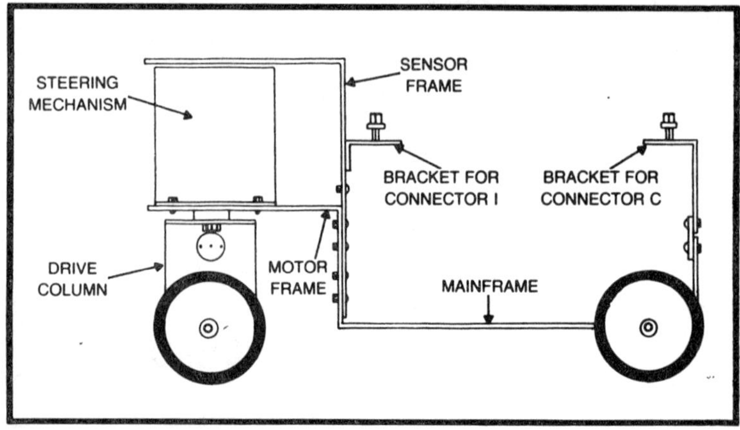

Fig. 3-1. The basic framework for the robot pet.

small and humble, so we knew our pet would have to follow suit. For this reason, the whole mobile cart had to be designed, from the ground up, as small and efficient as possible. This goal affected choice of motors, batteries, and overall mainframe configuration. As it turned out, height was not the limiting factor (except any that would affect stability), but rather width. A wheel-base of any greater than 12 inches would find it hard to move about freely between furniture and through doorways.

Second was travel reliability. The pet would not have a carefully-engineered environment of even tile surfaces on which to roll; there were rugs and perhaps small obstacles to contend with. And so the drive and steering mechanisms would have to be strong and stable. The present tricycle design was the result. A previous design was rejected, in which two independent front drive wheels and a rear free caster performed the task, much like a bulldozer. (Steering would have been accomplished by the separate energizing of either one of the two wheels.) Though this approach would have been simpler to construct and interface, it would have led to unpredictable motion under uneven loading conditions such as those carpets present.

The outgrowth of these deliberations was a narrow cart with one driven pivoting front wheel and two free-spinning rear wheels. The task then began with actual construction and a simultaneous search for motors and wheels.

The two main chassis parts are the *mainframe* and the *motor-frame*, shown in Figs. 3-2 and 3-3. The mainframe is merely a U-channel of ⅛-inch aluminum, with a width of five inches. (The width was selected on the basis of a tentative choice of batteries;

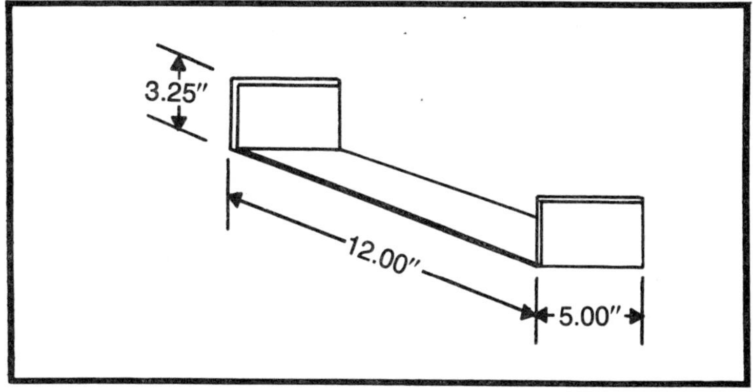

Fig. 3-2. Dimensions of the mainframe.

more on that later.) The motorframe is an L-bracket, also five inches in width, designed with the ultimate purpose of supporting the drive and steering subassemblies. The two frames were temporarily united by a 5″ by 6″ piece of aluminum (not shown), held with 6-32 machine screws and nuts, to facilitate design and early construction. (You may or may not choose to use such a plate, but it certainly helped me; when the drive and steering mechanisms were being added, it was a simple matter to place a weight in the U-channel to suspend the motorframe off a table for easy handling.)

CHOICE OF MOTORS

What sort of motors could be used? Basically I needed two 6-volt DC motors that were slow enough and had high enough torque

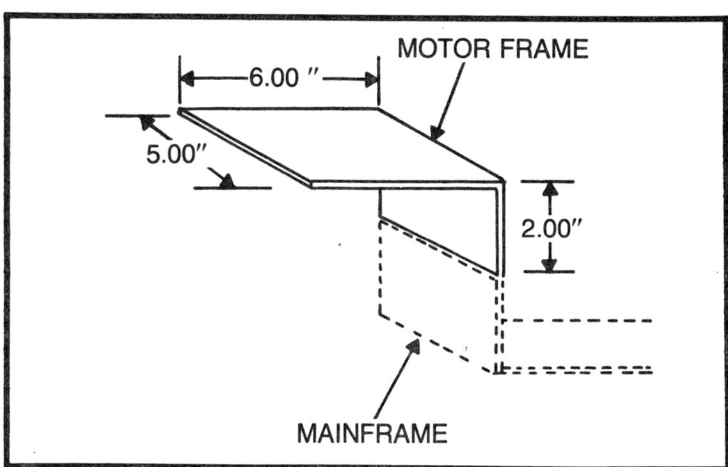

Fig. 3-3. Dimensions of the motorframe.

to pull the robot and turn it effectively. One approach is the use of *logic stepper motors;* these special motors rotate in increments of a few degrees at a time when properly pulsed. Speed, then, would simply be a matter of electronically controlling the frequency of a pulse train. Initial enthusiasm, though, gave way to some hard facts when I made some inquiries. For one thing, stepper motors are by no means inexpensive. The small market volume ensures a high cost. Secondly, steppers are not very efficient in terms of power consumption, which is a crucial consideration in a robot design with limited, available battery power. Finally, stepper motors are not very powerful per unit volume, and size is at a premium in a robot such as the robot pet.

The only way to go, then, is with some sort of 6-volt DC gearmotor. There are a number of companies that market these sort of motors, including the few listed in Appendix A, but we weren't sure what would work. So we tackled it in the piece-by-piece manner described herein. However, the reader is encouraged to find complete gearmotors which meet the specifications that resulted from our homebrew approach.

We broke the gearmotor question into its two parts: gearbox and motor. Many of the places which market complete gearmotors also market gearboxes alone. As it was, we used the gearboxes from a couple of defective rotisserie motors. The gearing was intact, but the AC motors were bad, so it was just a matter of removing the motor portions and leaving a shaft where the armature had been. One gearbox had a reduction of 440:1 and was perfect for the steering subassembly, which of course requires good torque and slow speeds. The other reducer had a ratio of about 60:1, which was fine for the main drive subassembly. The gearboxes were good, solid metal items, capable of reliable long-term use with nominal wear. Avoid the cheaper plastic gearboxes if possible.

Motors are next in line. One excellent source for DC motors is a defective cassette or eight-track cartridge player. The motors built into these tend to be small and yet efficient, making them desirable for the size-stingy robot pet. Motors of this kind, however, often have some sort of speed governor built into their design. A standard speed governor operates by shutting down the motor whenever it rises above a desired rpm. For robotic applications, this is an inefficiency which cannot be tolerated. It amounts to duty-cycle-powering the motor, and we need the most power possible from the motor. So, in such motors, it is necessary first to defeat the governor.

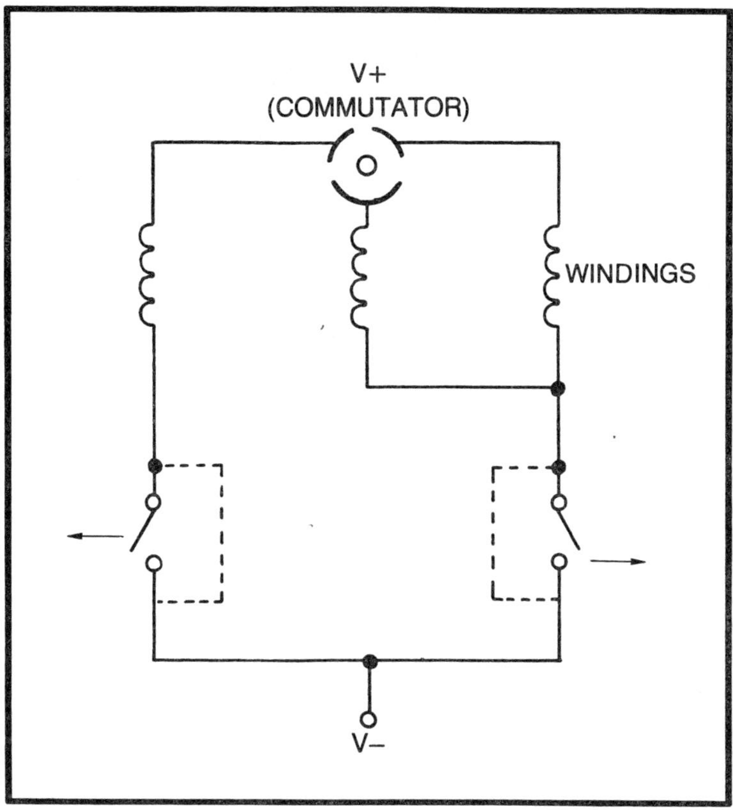

Fig. 3-4. Standard motor speed governor. The governor is defeated by the addition of the jumpers indicated in dashed lines.

Figure 3-4 gives the schematic for a standard governor. As the rpm increases, the switch contacts open, causing the motor to shut down for a fraction of a second. All that need be done is to short the switches with a piece of jumper-wire, as shown in dashed lines. In this way, the switches are bypassed, and current to the windings is maintained. When possible, it is advantageous to remove the whole switches and solder the winding leads directly together. This reduces the mass of the armature a bit and increases motor efficiency correspondingly.

How powerful a motor do you need? That depends on a few factors. First, we have power limitations based on the batteries chosen. The robot pet uses two motorcycle batteries for a supply of 6 volts at 24 amp-hours. That is to say, the supply will source 1 ampere for 24 hours, 2 amps for 12 hours, 3 amps for 8 hours, and so

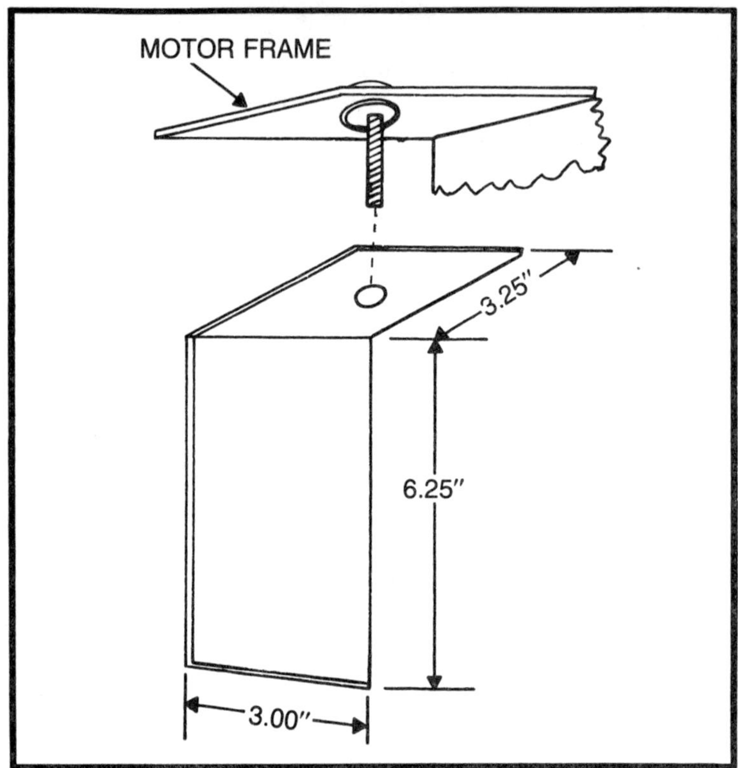

Fig. 3-5. Dimensions of the drive column.

on. And so, if we intend for the robot pet to have a reasonable longevity between charge-times, we cannot have drive and steering motors that draw 10 amps each. The motors should draw about an amp or two each under fully-loaded conditions.

Second, consider speed. The robot pet need not gallop—indeed, most of us are upset when the family dog runs around too much. The best average speed is somewhere between ½ and 1½ miles per hour. Assuming a six-inch diameter wheel, the front drive-shaft would have to rotate at about 60 rpm to average one mph. Using a gear-reducer of 60:1 ratio, this implies a DC motor of about 3600 rpm. For the steering column, a speed of about eight rpm is optimum. Using the 440:1 reduction, another DC motor of about 3600 rpm fills the bill.

In the choice of motor speeds and gear reductions, do not be afraid to run things a bit on the slow side. The robot pet will weigh in at about 20 pounds when all is through. It will require some torque to move such a mass, taking obstructions such as carpet into consideration. Better to increase the torque at the expense of overall speed than to have a pet who is forever getting stuck on the edges of rugs!

MOUNTING THE MOTORS

Whatever the choice for a main drive gearmotor, it is mounted on a L-bracket as shown in Fig. 3-5. The L-bracket can be pivoted at a point directly above the front wheel, using the set-up shown in Fig. 3-6. A ball-bearing assembly of 1.75-inch outer diameter and 1.00-inch inner diameter is mounted on the motorframe bracket, above a 1.25-inch diameter hole. In this way, a ½-inch bolt can be placed through the bearing, and coupled directly to the lower L-bracket. This allows the entire drive gearmotor assembly to rotate freely beneath the motorframe. Then it is a matter of simply coupling a second gearmotor to the ½-inch bolt from above, to provide steering.

In the prototype pet robot, the steering gearmotor was mounted on two brackets directly above the ½-inch coupling bolt which formed the pivot-point for the drive column. The output shaft of the gearmotor was 3/16-inch, and so a hole was actually drilled

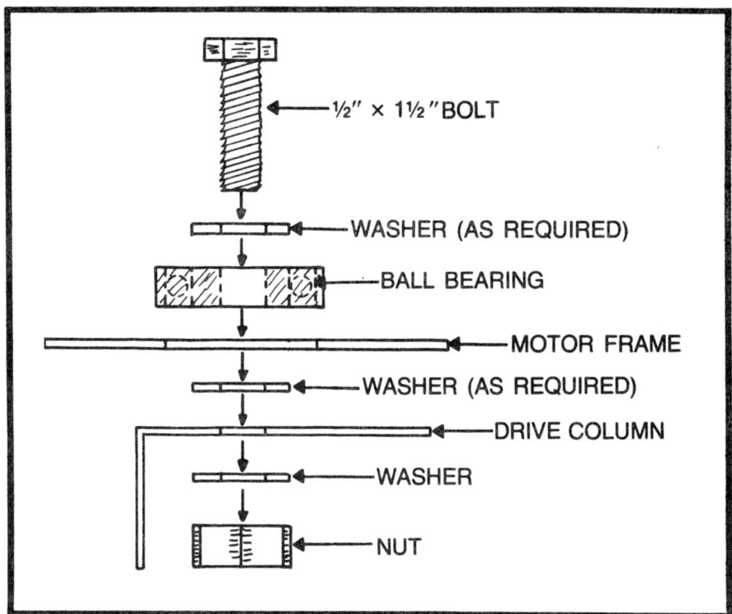

Fig. 3-6. Coupling the drive column to the motorframe.

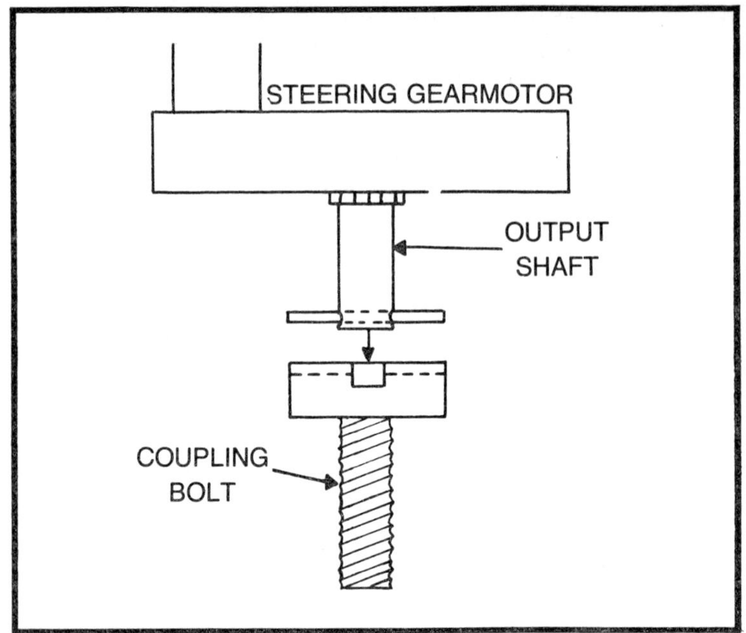

Fig. 3-7. Coupling the steering gearmotor to the drive column.

part-way into the head of the ½-inch bolt, permitting the output shaft to fit snugly into it, as depicted in Fig. 3-7. To complete the coupling, a groove was cut into the head of the bolt, somewhat shallower than the depth of the hole previously drilled for the gearmotor shaft. Then a hole was drilled and an ⅛-inch pin placed through the gearmotor shaft, such that the pin fits comfortably into the shallow groove in the ½-inch bolt head. So it was a simple coupling, one which permitted easily removing the gearmotor for purposes of further construction or improvement.

In both cases of gearmotors, the prototype made use of separate gearboxes and motors. The steering motor came from an eight-track cartridge player, modified as described earlier so that the speed was not governed internally. The drive motor had to be quite a bit more powerful; it was a surplus DC motor with a current-draw of about 2 amps maximum under loaded conditions. Both motors were mounted next to their respective gearboxes, with output shafts parallel to the input shafts of the gearboxes.

As was explained earlier, the gearboxes had originally been driven by AC motors, which we removed, leaving only the armature. The armature easily pressed off of its central shaft, but then there was the difficulty of coupling the shaft to the output shaft of the DC motor. We pressed a pulley onto each shaft and used a neoprene O-ring to complete the coupling. The pulleys were homemade out of aluminum and such, but certain standard-sized pulleys are available at various hardware shops.

The completed drive and steering mechanisms appear as in Figs. 3-8 and 3-9. The reader's version may differ in minor details,

Fig. 3-8. The completed steering mechanism.

Fig 3-9. The completed drive column.

but the overall concept of the swiveling bracket and geared drive is what counts. You may wish to add a few supportive brackets or braces to reinforce the motorframe; after all, much stress will be placed at the point where the mainframe and motorframe meet. The sturdier the construction, the better and more enduring the pet robot will be.

WHEELS

The final touch to this stage of body construction is the addition of wheels. The best for the job are six-inch diameter, rubber-treaded wheels, like those used on lawnmowers or small carts. Most garden supply stores or hardware stores should be able to furnish these for a few dollars each. There are two kinds needed: those with built-in ball-bearings, and those with stationary sleeve bearings. The robot pet uses two free-spinning wheels in the rear, for which the ball-bearing version is ideal. The front drive wheel, on the other hand, couples directly to the drive gearmotor output shaft, and thus the sleeve-bearing version is best for this one.

There are a number of ways to couple the front wheel to the drive shaft, depending on the particular gearbox and wheel. In the prototype, the gearbox shaft was not long enough to reach through the wheel, and thus a shaft extender had to be made. (The inner dimension of the extender fit the drive shaft, and the outer dimension fit the wheel.) Two small brackets were mounted on the outside of the drive wheel, through which a long 4-40 machine screw could extend. This same 4-40 screw also extended through a hole in the end of the shaft-extender (which protruded out from the wheel's sleeve bearing), thus effectively coupling the two.

For the two rear wheels, two right-angle brackets were fabricated from steel, each appearing as in Fig. 3-10. On each bracket, a ½-inch bolt was welded to act as a "dead axle" for the built-in ball-bearings of the rear wheels. (In the absence of welding tools, of course, a ½-inch bolt could simply be mounted *through* the right-angle bracket.) The ends of the ½-inch bolts were drilled to permit the insertion of cotter pins to hold the wheels in place. The wheels were placed on the bolts, and the ball-bearings were held stationary by a nut on the outside. (A hole drilled through this nut allowed a

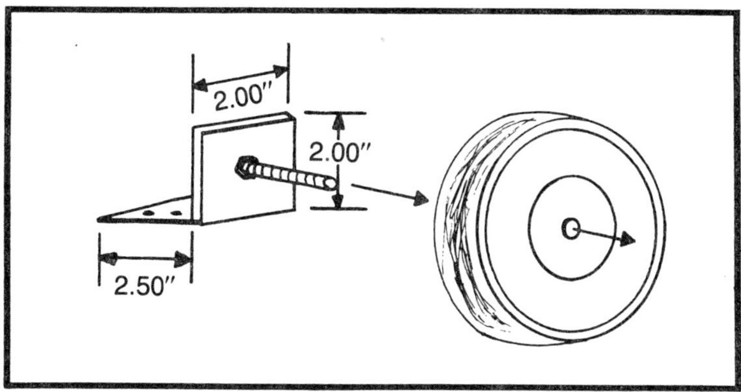

Fig. 3-10. Dimensions of one rear wheel bracket.

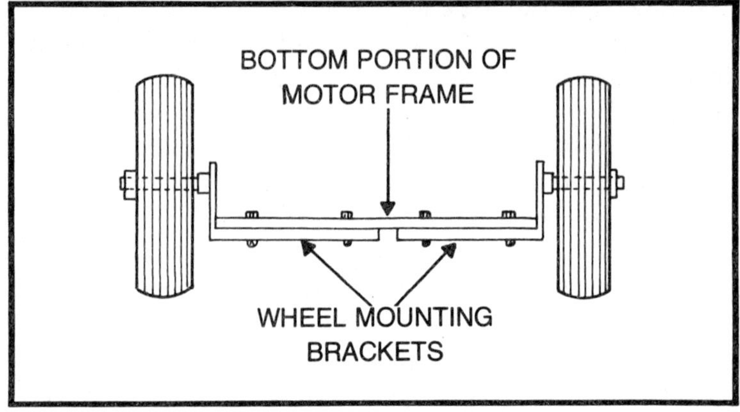

Fig. 3-11. Mounting the rear wheel brackets to the mainframe.

cotter pin to extend through the corresponding hole in the bolt, keeping the nut in place.) With the bearings stationary, the wheels could turn freely, permitting the chassis to roll. Figure 3-11 depicts how the two rear-wheel brackets mount to the mainframe. The fact that the two wheels rotate independently, rather than in tandem, contributes to the ease of steering of the robot pet in motion.

FINAL TEST

Before moving on to the next chapter, which describes further additions to the chassis, it is wise to put the pet robot through its paces. Set the chassis down on a clear, wide space on the floor. Using a 6-volt source capable of an amp or two of current, apply power to the steering gearmotor. Observe the motion of the steering column, as you operate the motor in either polarity. Does anything bind or put an unexpected load on the motor? Check the couplings and the main bearing for any obstructions.

Next, apply power to the main drive gearmotor—forward, then reverse. Does the chassis move freely? Can it negotiate small obstacles such as small sticks, or carpet edges? Again, check the bearings and couplings for any sign of binding. Remember that if the chassis only nominally navigates now, it will never make it with an extra ten to fifteen pounds of equipment on its back. Design for *power*, not speed.

Chapter 4 describes the addition of some important parts to the chassis, parts which relate to the control of the drive and steering by the brain board (still a ways off). But before you leap into the next chapter, take a moment to appreciate how far you've come. True, that little motorized tricycle is nothing in comparison to the robot pet in final form. If you've gotten this far, though, the remainder will not be that difficult to complete. Chapter 4 will show you how to prepare for that long-awaited day when the chassis will simply get up and roll away—quite on its own.

Chapter 4
Interfacing The Body

Imagine, if you can, sitting in a brand new automobile, with a souped-up engine, a good set of tires, and a full tank of gas. Then imagine the frustration when you realize that the car has no steering wheel, no ignition, and no accelerator!

That's about the status of the pet's chassis right about now. You could take a small computer, if you wished, and place it in the back of the mainframe, and the whole contraption would just sit there. The problem is self-evident; there must be some way for the computer in the back seat to control all of that locomotory power up front. Putting it plainly, there must be a means of *interface*.

SELECTION OF RELAYS

So far, we can take a 6-volt source and hook it directly to the gearmotor wires, and get the chassis to move. But how can a computer do something like that? The computer needs some way to convert its low-power signals into true power control. It has to be able, not only to switch the battery voltage to either or both gearmotors, but also to switch polarity to those motors to obtain opposite directions. The easiest way to do this is with relays.

Not just any relay will do. First of all, it has to have contacts that are rated at a couple of amps, to handle the current required by the drive and steering motors. Second, to handle the polarity-switching aspect, it has to be double-pole, double-throw. And third, it has to be an *efficient* relay; with power at a premium, the pet robot cannot

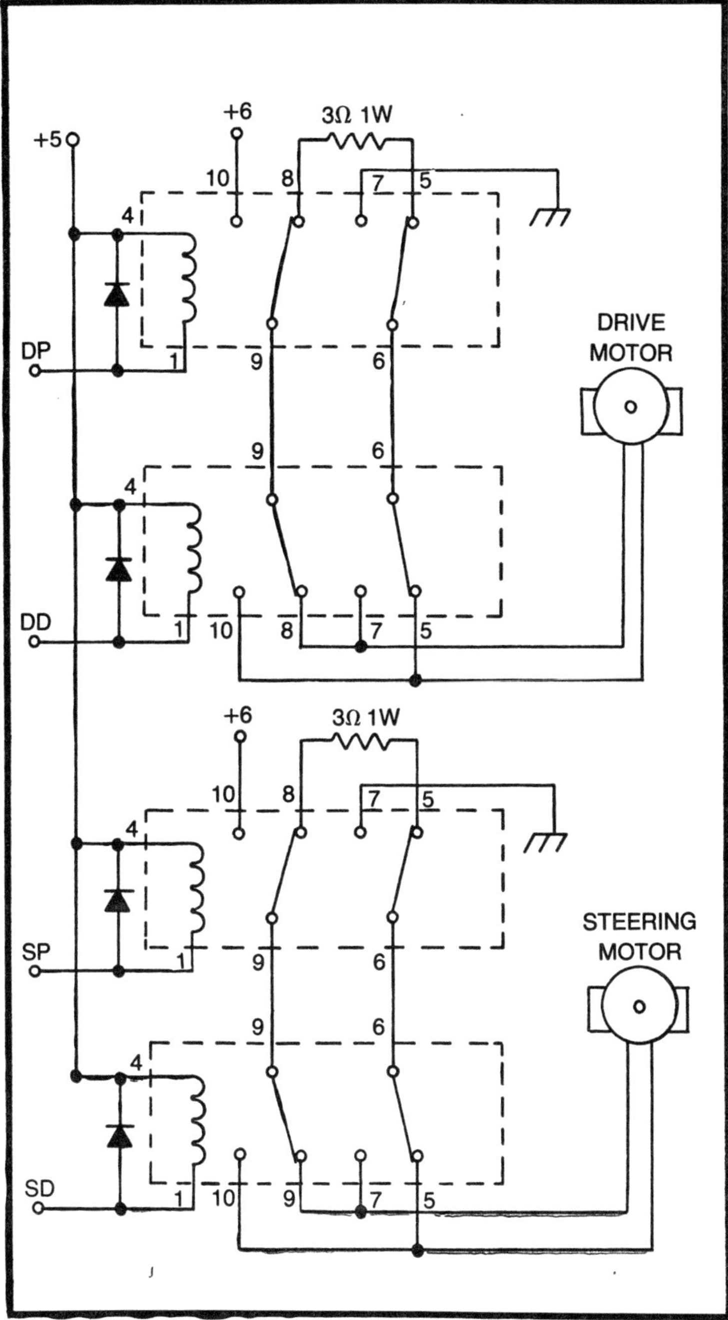

Fig. 4-1. Locomotory control relays. Each of the four relays pictured is Potter & Brumfield R10S-EI-Y2-J1.OK; diodes shown are 1N4003.

afford to burn up 6 volts at 200 milliamps per relay. If at all possible, it is desirable to use a relay which could be driven by a TTL buffer integrated circuit—that is, requiring something on the order of 20 milliamps at the maximum.

That is a tall order to fill, but there *are* relays that perform this efficiently. The Potter & Brumfield relay specified in Fig. 4-1 is no doubt only one of many offered by relay manufacturers nationwide that operate as low in current-consumption as required. The R10S relay used in the prototype pet robot operates at 5 volts at 5 milliamps, quite easily driven by the proper TTL integrated circuit on the brain board. If these relays are too difficult to obtain, you might substitute standard high-current relays, but be prepared for two things: shorter battery life, and the need for driver transistors. (Standard 2N2222 NPN transistors may be used as drivers in such a case, but take note that this inverts the logic levels required to activate the relays.)

Figure 4-1 shows the wiring of the relays. One relay for each motor is required for an on/off switch; another relay for each motor reverses the polarity to the motor for a forward/reverse function. There are four relays, then: the drive power relay (DP), the drive direction relay (DD), the steer power relay (SP), and the steer direction relay (SD). Remembering that the currents involved may be in the one- or two-amp range, be sure to use fairly heavy wire to connect the motors, say 20-gauge stranded, insulated wire. Smaller wire gauges will lose power as heat, and reduce the effective output of the gearmotors.

You will notice that the on/off relays (DP and SP) in both cases choose to hook the motor to one of two places: either the 6-volt supply or a shorting resistor. The shorting resistor effectively shorts out the back-emf generated by the spinning motor, causing the motor to jerk to a halt. The end result is *reduced overshoot,* which is particularly important in the steering subassembly. The resistances have been selected so as to limit the "spike" of shorting current to 2 amps, which is the rated contact current specification of the relays.

One important factor that Fig. 4-1 does not specify is the *direction* of the motors with relation to the relays. This is particularly crucial in the case of the Servodrive steering control. The drive motor should be wired so that it runs *forward* when the DD relay is *not* energized, and *right* when the relay is energized. The interplay of the four relays will be treated in detail in Chapter 8.

One side of each relay coil is tied to the +5-volt supply of the finished pet robot system, and the other side is left floating. All the brain board then needs to do, to energize a given relay, is to pull the floating side to ground sufficiently to sink 5 mA. (These control-points will be referred to by the two-letter designations shown in the figure.) Notice that 1N4003 diodes are placed across the relay windings. These help protect the brain board from any inductive kickback or "spike" which might be generated by the opening or closing of the relay.

MOUNTING THE RELAYS

The sockets for the R10S relays can be mounted side by side, taking up very little space. The four relays in the pet robot prototype are mounted upside-down on a new chassis angle which is to become the platform for the "head"of the pet. Before obtaining the relays, then, this bracket can be added.

Figure 4-2 shows the new L-bracket, which might be termed the *Sensorframe,* since most of the pet's sensory organs will eventually be mounted upon it. This Sensorframe takes the place of the temporary mounting-plate which originally held the mainframe and motorframe together in the last chapter. It would help speed up things if you used the old temporary plate to aid in the marking and drilling of any mounting holes already used.

The four relays can be mounted upside-down, beneath the new sensorframe. Since this is a difficult place to reach once the new

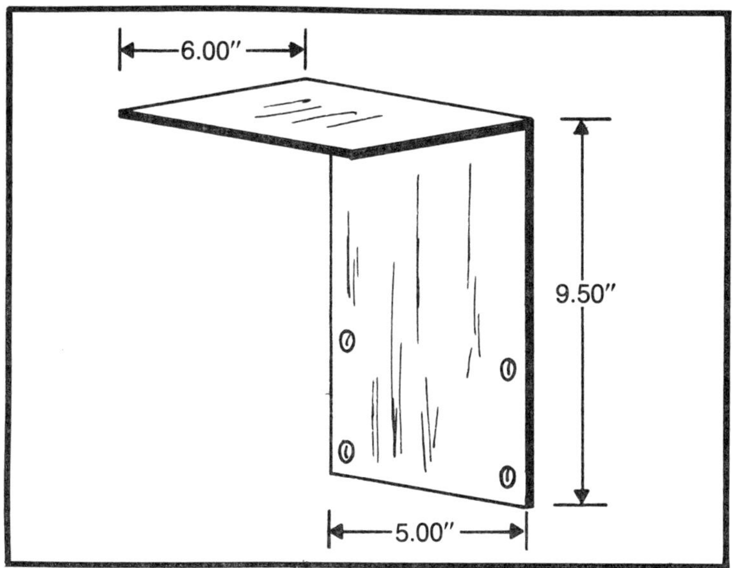

Fig. 4-2. Dimensions of sensorframe.

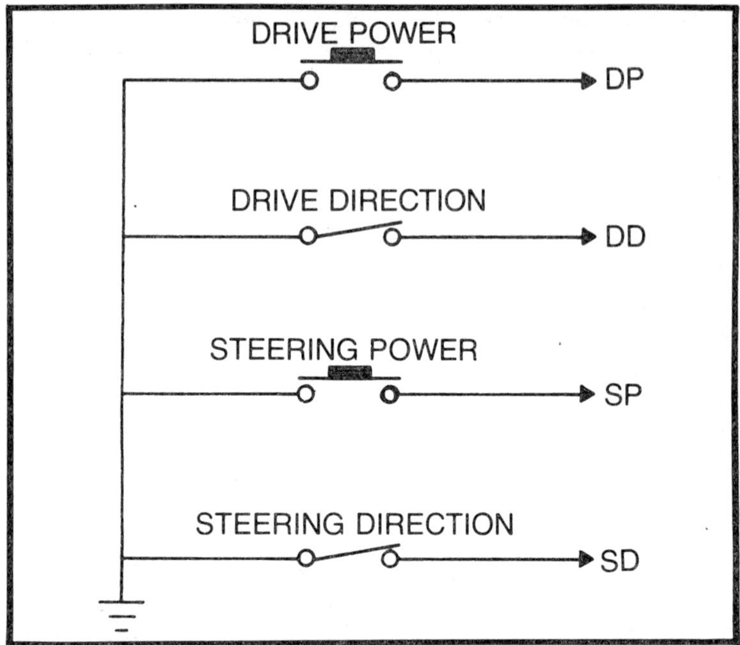

Fig. 4-3. Optional manual control link for locomotory system.

frame is mounted, it is a good move to mount the relay sockets to some sort of unifying bar or plate, and *then* mount that piece to the sensorframe. Be careful, of course, that none of the socket connections touch the bare surface of the frame.

Once the four relays are wired and mounted as directed, you may wish to construct a simple manual control link, as shown in Fig. 4-3, just to get the "feel" of the robot pet's locomotory system. The controller is a simple bank of four switches, intended merely to sink any one of the four relay windings to ground. Since the relays only draw 5 mA each, the leads from the controller to the pet need only be light gauge wire, such as 26-gauge. Run power leads from an external power supply to provide the 6 volts necessary to drive the motors, and a +5 supply voltage to the indicated pins of the relays. Of course, in lieu of the necessary supplies, the reader may hold off on building the manual link until the pet's own supply is designed and placed "on board." The subject of the battery supply is handled in detail in Chapter 5.

SERVO POTENTIOMETER

The brain board now has a responsive motorframe awaiting its command. One more important factor in the motorframe need be added, and that is the *servo potentiometer*. This added part will provide the necessary feedback loop to enure reliable steering capabilty.

The need for the servo pot arises from the necessity of monitoring and altering the steering column angle. This sort of thing can be handled in a number of ways, from simple microswitch closures to high-resolution stepper-motor systems. The robot pet makes use of a medium-complexity feedback circuit termed the *servodrive*, which is described completely in Chapter 8. In order to ascertain the present position of the steering column, the servodrive requires a reference voltage, based on the steering angle. In this way, the voltage obtained from the column can be compared to a control voltage introduced by the microprocessor and course corrections can be initiated.

The best method of providing such a reference voltage is to couple a potentiometer to the steering column, as shown in Fig. 4-4. As the column rotates from a full left turn to a full right turn, the potentiometer wiper follows. By placing 5 volts across the potentiometer, an output voltage of from 0 to 5 volts is obtained. (Later, when the servodrive circuit is actually built, the resistors shown in Fig. 4-4 will be selected and added. Their purpose is to narrow the reference voltage range to a 1-volt-to-4-volt sweep.)

Since the coupling for the steering gearmotor allows the steering assembly to be removed somewhat readily from the actual column, it is a good approach to maintain the "modular design" by

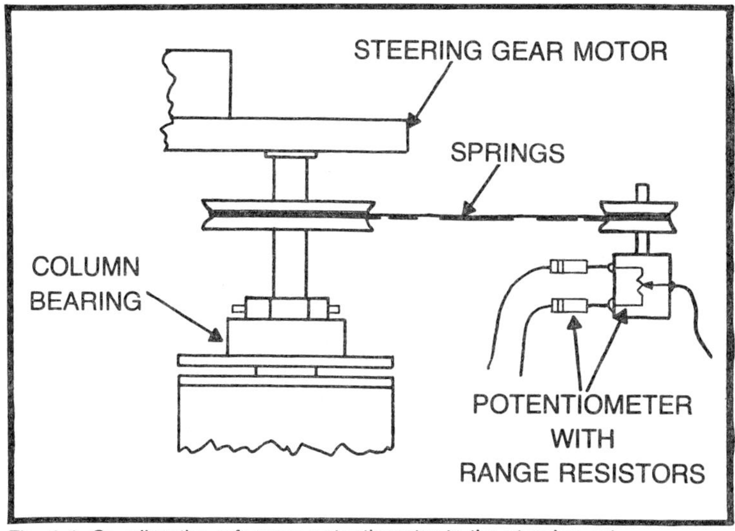

Fig. 4-4. Coupling the reference potentiometer to the steering column.

mounting the servo potentiometer as a part of the steering assembly. Otherwise, it is much more difficult too remove the steering motor for repair or improvement without "tangling up" the linkage to the potentiometer.

Choose a good potentiometer of linear-taper characteristics and single-turn design. The resistance should be at least 10 kilohms, or preferably 25 K, as was used in the prototype. The higher the resistance the better, since the resulting reference voltage will not be loaded at all by the Servodrive circuit, so there is no need to add to the load on the +5 supply by drawing excess current through the potentiometer. A 10 K pot will draw only ½-milliamp, and a 25 K pot will draw a mere fifth of a milliamp.

Place a pulley on the output shaft of the gearmotor as shown in the figure and a second pulley on the potentiometer shaft. The ratio of these will have to be such that the potentiometer shaft will make one full rotation (or whatever percentage of full-circle it takes to sweep through its range) for one complete left-to-right arc of the steering column. One way to arrive at a good estimate is as follows. Decide on the maximum steering angles to be involved; the column should be free to turn a maximum of about 70 degrees right or left of center, or a total arc of 140 degrees. Next, determine the full left-to-right sweep of the potentiometer; it is probably on the order of 300 degrees. Thus, the ratio of the circumferences of the pulleys is 140/300, or about 1:2.14. And since diameter is a factor of circumference, the ratio of 1:2.14 holds for the diameters as well. In short, the pulley on the steering column must have about *twice* the diameter of the pulley on the potentiometer.

Don't let the numbers frighten you. If your particular set of parts yields the angles given above, you need not search all over creation for a pulley exactly 2.14 times the size of another. Twice the diameter will do; it simply means that the entire sweep of the pot will not be utilized, a loss which is quite liveable. But if the ratio were too large—say you chose a pulley 2½ times the other—the pot would run out of room before the column had completed its maximum sweep, which would result in areas of imprecision under the control of the Servodrive circuit. Feel free, then, to approximate and experiment, but do so in the direction that does not slam the pot up against the internal stops.

To couple the two pulleys, we could use a neoprene O-ring, as in the drive assemblies, but in time that would allow slippage and gradual inaccuracy. Instead, use a string, or some light fishing line. The line is wrapped a couple of times around each pulley to prevent any slippage. Two small springs are incorporated into both sides of the loop, to keep the line taut and to prevent damage or breakage of the line should the potentiometer reach its limit. Of course, usage of the springs requires sufficient length of line on either side of them, so that the springs themselves do not actually roll up onto either pulley. Couple the pulleys while both the gearmotor shaft and the potentiometer are centered.

To test the arrangement, replace the steering assembly with the new potentiometer onto the steering column. Then energize the steering motor, and turn the column fully to the left, then to the right. If the column turns too far, one of the in-line springs should be seen to stretch some, protecting the potentiometer and the coupling line. If a 5-volt supply is hooked to the stationary contacts of the pot, the wiper contact should output a clean zero-to-5-volt sweep as the column makes its complete left-to-right turn.

CONNECTORS

One final addition will be made to the chassis at this stage, an important addition as far as the brain board is concerned. This is the adding of *connectors* to permit the plugging-in of the brain board. The beauty of the connector philosophy is *modularity*; the brain board can be removed at any time for easy inspection, repair, or expansion. The connectors must be purchased, true, and it is a chore to wire them up properly, but the simplicity they afford far outweighs the extra burden. In fact, their use reinforces a good healthy discipline towards neat, careful wiring and the thinking-through of interconnection.

The connectors used are of the DB-25 type, a commonly available connector through most electronics mailorder companies or local computer shops. The chassis holds two DB-25S connectors, the female version, which receive the male DB-25P connectors to be mounted later onto the brain board. Mounted at the front and rear of the mainframe, the DB-25S sockets allow the brain board to plug in directly above the battery supply (described later). In addition, a DB-25P male plug is mounted in the bottom or rear of the mainframe, and this plug becomes the means by which the owner can externally access the brain board for programming.

If you consult Fig. 3-1 of the previous chapter, you can see the manner in which the two upper connectors are mounted. The frontmost connector, termed connector "I" for "interface," is placed on a small angle-bracket which is then in turn bolted onto the sensorframe L-bracket added earlier in this chapter. Note that the connector is actually "standing up" on a short pair of standoffs or spacers. This saves the trouble of having to cut and file the unusual

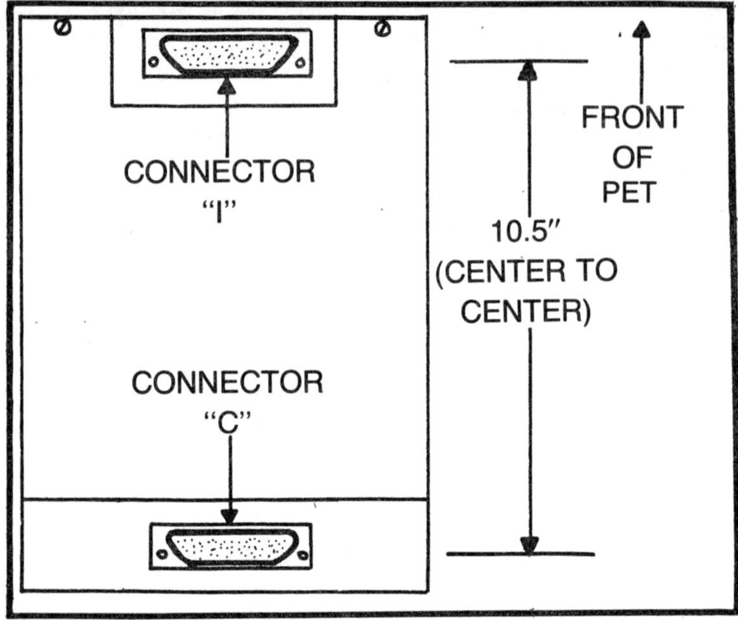

Fig. 4-5. Orientation of DB-25 connectors.

trapezoidal hole it would take to mount the connector through the bracket. Plus, it is fairly easy to remove the connector and reach the solder connections underneath; no wires are fed irremovably through a mounting hole.

The rearmost connector, termed connector "C" for "control," is mounted also to a somewhat larger angle-bracket, this piece being the full width of the mainframe. The new angle-bracket unites with the mainframe, the two being held together with a mounting-plate. (In the prototype, the mounting-plate used is the temporary plate which was displaced by the sensorframe earlier in the chapter.) The connector is mounted on standoffs again, and should be on a horizontal level with the frontmost connector. For this reason, it is probably best to mount the rear connector first, then select and drill mounting holes for the front connector bracket as required.

One all-important consideration is that the measurements given for the positioning of the connectors may need some adjustment, depending on how closely the reader has held to the mainframe measurements earlier in the book. The prime measurement is that the two connectors must be 10.5 inches apart, center to center. This is to allow connection to the corresponding brain board connectors, which will also be mounted 10.5 inches apart. The bracket measurements should provide the approximate distance, but the reader may want to slot the connector mounting holes for exact adjustment later.

Note also the orientation of the connectors as shown in Fig. 4-5. For safety, the connectors were deliberately mounted with the pins oriented the same for both connectors. Had the orientation of the connectors been centrally symmetrical, it would have been possible to plug the brain board in backwards. The shape of the DB-25's as they are mounted in the prototype prevents such board reversal and the subsequent board damage which would almost certainly occur.

Next is the mounting of the external access connector, a DB-25P. This connector, termed connector "A" for "Access," may be mounted in the bottom of the chassis, or the lower rear, depending on the amount of space. In any case, a trapezoidal hole needs to be cut and filed for it; the connector is then mounted from the inside out, such that any connecting wires do not feed through the mounting hole as such. Figure 4-6 shows the positional relationship of connector "A," the recharge jack, and the reset switch, the latter two of which will be added in later chapters.

Prior to actually mounting the external access connector, it should be wired to its "partner," which is the rearmost upper DB-25S. The wiring configuration is one-to-one , pin-for-pin. Measure the approximate length needed to route the 25 leads down from the upper bracket to the lower connector mounting location. Then cut 25 identical pieces of insulated wire, about 26-gauge, and connect the two DB-25s on the workbench, where there is plenty of room. Then mount them back where they belong.

The end result of using the lower DB-25P connector and wiring it as described is an externally-available connector identical in type and wiring configuration to the corresponding plug on the brain board. Later, when the manual programmer is built, this will prove to be an advantage; the programmer can plug into the access connec-

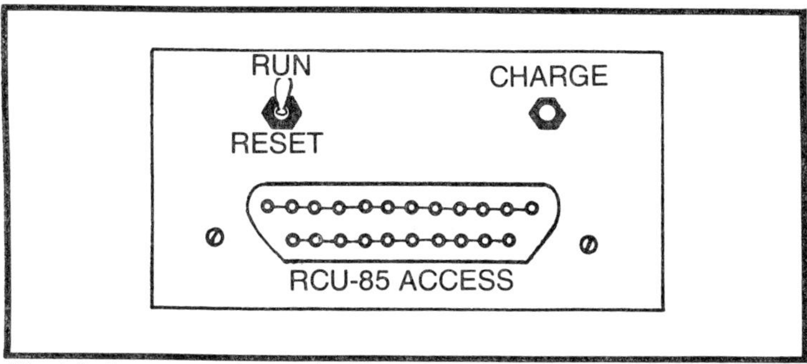

Fig. 4-6. Access panel on underside of the robot pet.

Table 4-1. Pin Configurations For DB-25 Connectors

Pin	CONNECTOR "I"		Pin	CONNECTOR "C"/"A"
1	+5 V		1	+5
2	+6		2	
3			3	CLK (OUT)
4	DP Relay	Servodrive Signals	4	$\overline{\text{RES IN}}$
5	DD Relay		5	IO/$\overline{\text{M}}$
6	SP Relay		6	$\overline{\text{RD}}$
7	SD Relay		7	$\overline{\text{WR}}$
8	Feedback V		8	ALE
9			9	SID
10			10	SOD
11	Transducer L_i	Soniscan Inputs	11	
12	Transducer C_i		12	
13	Transducer R_i		13	
14	Transducer R_o	Soniscan Outputs	14	AD_0
15	Transducer C_o		15	AD_1
16	Transducer L_o		16	AD_2
17			17	AD_3
18	Excom Mike		18	AD_4
19			19	AD_5
20	Audigen Output		20	AD_6
21			21	AD_7
22			22	A_{14}
23			23	A_{15}
24			24	
25	GROUND		25	GROUND

tor, or it can connect directly to the brain board plug while the board is separate from the chassis.

Up until now, connector "I" has merely been mounted and not wired. This is because very little of what the connector is used for has yet been mounted. There are, however, a few things that now can be wired to this connector. Specifically, these are the four control leads to the servodrive relays and the one lead to the wiper of the servo potentiometer.

Table 4-1 gives the pin configuration for the two DB-25S connectors. Most of the designations do not mean much right now, but this chart will be referred to from time to time later in the book. Of special attention at the moment are pins 4 through 8 of connector "I," which relate to the servodrive circuit. Run leads from the designated relays and the potentiometer to the pins, either around the edge of the sensorframe or through a hole in the frame. Since all five levels are very low current, use a light-gauge insulated wire as on other occasions, such as 26-gauge. Then remount the connector as previously. (Do not be concerned, by the way, that two of the three potentiometer terminals are presently unconnected; these will be dealt with in a later chapter, when the two resistors shown in Fig. 4-4 are added.)

Little by little, the robot pet chassis has grown. The additions prescribed in this chapter, except perhaps for the relays, have done little to improve the visible functioning of the chassis. And yet, as future chapters will clarify, these additions have paved the way for the *ultimate improvement*—the brain board. Before we can start the brain board, though, we must make a necessary detour. For the robot pet must be truly mobile, and it can never be so, until it has its own independent on-board power supply. The details of this crucial development, from selection of a battery to construction of a charging system, are handled one by one in the next two chapters.

Chapter 5
The Power System

The robot pet, once completed, will be a pet geared for activity. It will be doing some complex calculation and some impressive moving about. This being the case, it will require, like its living cousin, some form of nutrition, a source of energy. Dogs can eat food (if we provide it), but the robot pet wouldn't know what to do with a bowl of beef if it had one.

The energy problem could be solved for the robot pet if we chose to keep him "leashed" to a line-powered supply, but a pet needs freedom to roam. So the problem of power leads invariably to batteries.

Types of batteries abound, and modern-day technology manages to make breakthroughs in battery efficiency regularly. But we have various limitations on our choice, including such things as size, weight, voltage, current capability, charge characteristics, and cost. None of these factors can be lightly overlooked in making a selection of battery.

As to the actual battery type, a few exclusions can be made from the beginning. Dry cells, for instance, are lightweight and inexpensive, and there is no danger of spilling corrosive electrolyte from them. But dry cells are usually not rechargeable and not very powerful. (Remember, after all, that the motors alone will require a couple of amps of current.) Nickel-cadmium cells are rechargeable but require a very careful recharge cycle. Plus they are somewhat costly. Sealed "gel cell" batteries are a step in the right direction, being clean and rechargeable. But they do not often come in a variety with a high enough ampere-hour rating. And the cost is still uncomfortable. All things considered, then, the final and best alternative to date is the familiar, old-fashioned lead-acid wet cell, with gel cells as a good possible second design choice.

VOLTAGE REQUIREMENTS

Next is a question of voltages. What specific voltages will we need, and how can we configure a battery system to handle them all? Certainly, for TTL integrated circuits, a solid +5-volt supply is imperative. And depending on the selection of motors, a 6-volt or 12-volt supply will be required. If any op amps are utilized in the control circuitry (and there certainly are in the robot pet), then ordinarily we would need a bipolar supply, say a plus-and-minus 5-volt one. With a little thought and experimentation, though, this can be gotten around, using op amps in a single-supply manner, and thus simplifying things a bit.

Now, let's suppose we use a 12-volt motorcycle battery for the basic power source. That's fine as far as the motors go (if 12-volt motors are selected), and most op amps are built to run at such a voltage or greater. But the TTL circuitry, and notably the Intel 8085A microprocessor, demands a reliable +5-volt supply.

There are a number of excellent single-package *voltage regulators* on the market, ranging from simple fixed-voltage versions (the 5-volt LM309K) to variable precision regulators (like the popular 723). Any one of these could be instituted to regulate the 12 volts of the battery down to a stable 1 ampere, +5-volt supply. The only problem with such an approach, though, is sheer inefficiency. These standard regulators are resistive in nature; that is, they regulate by resistively "burning off" any unwanted voltage. Consider the waste, then, of burning 12 volts down to 5 volts. It amounts, literally, to producing more heat from the battery output than usable power for circuitry.

You may answer quickly, "Then that's why you've chosen to use a 6-volt battery instead...so that there would be less waste by the regulator." Do not answer too quickly, however, for there is a catch. Most solid-state regulators have *inherent* inefficiencies which limit their usage. As a matter of fact, the average +5-volt regulator requires an extra 3 volts just because of the limitations of its internal transistors. To put it plainly, the average regulator needs at least 8 volts to supply an output of 5 volts. A 6-volt battery just doesn't cut it.

That's not the final word, though. All would be fine if a circuit existed that *could* regulate 6 volts to 5. And, fortunately, such a circuit does exist, and details will follow shortly. It is sufficient to say at this point that, given such a hypothetical "high-efficiency" regulator, the best battery configuration is a hefty 6-volt motorcycle battery.

As an aside, you will notice that relays were used in the motorframe of Chapter 4 to accomplish the necessary reversal of the drive and steering motors. Perhaps you thought at the time that this was an amateurish approach and somewhat of a step backward from the "state-of-the-art" technology promoted in the rest of this volume. After all, if you are at all familiar with standard servomotor systems, it is quite the standard practice to energize motors with power transistors, using a bipolar supply to provide the forward and reverse potentials. But, once again, simplicity in the power supply won me over; if relays could save me from the complications of a bipolar supply, why not use relays, archaic as they may seem? Plus the use of relays afforded another little advantage: the ability to eliminate motor overshoot by "shorting" the windings when not energized. And finally, seeing that I was planning on 6-volt motors, I was voltage-stingy; a closed relay is essentially a resistless pathway, but a saturated transistor may burn off a good three-quarters of a volt! The battle cry of "Efficiency" won out again.

CURRENT REQUIREMENTS

Batteries such as those used in the robot pet usually come specified with an *ampere-hour rating*. A battery with an amp-hour rating of 12 can source 1 ampere for 12 hours, 2 amperes for 6 hours, and so on. Beyond the time limit, the voltage begins to drop. A 6-volt, 12-amp-hour battery will finally drop to about 5 volts after 12 hours of sourcing 1 amp, and beyond that, it falls rather quickly into "deep discharge." It's easy to see, then, that the choice of amp-hour capacity directly limits the longevity of the robot pet, at least as it relates to duration between charging periods.

The motors as chosen should not draw more than an amp or two, and they are energized intermittently, not continuously. So it might be a fair estimate to think of the motors as requiring the equivalent of 1 amp continuous. The electronics on the brain board will require about 1-amp continuous. Therefore, we can expect an estimated demand of about 2 amps continuous on the power supply. From there, the robot pet's longevity becomes a simple problem in algebra. With a 12-amp-hour battery, the pet would last 12/2, or six hours, between charges.

Personally, for the prototype, I wanted a pet with a little less "hunger." So I set about to find a 6-volt motorcycle battery with at

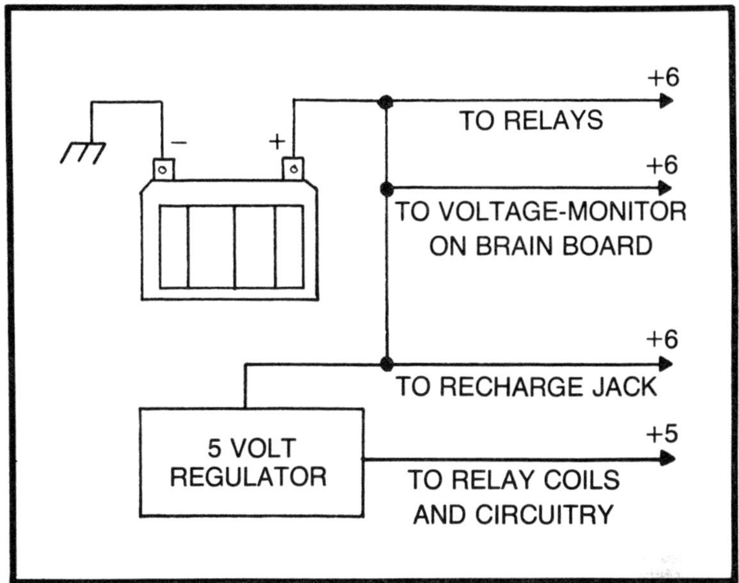

Fig. 5-1. Output distribution of the power system.

least 18 or even 24 amp-hours of capacity, the idea being that the robot pet could be left to run all day, if desired, and then be recharged in the evening. The only problem was *size*. There were few 6-volt 24-amp-hour motorcycle batteries to be found, and those I found were too tall for the already-constructed mainframe. The only thing left to do, as painful as it was to my wallet, was to purchase two 12-amp-hour batteries and connect them in parallel. The two batteries were small, German-made models; their height was about five inches and their width, when added together, was a comfortable four and three-quarter inches, just right for the five-inch width of the mainframe. Of course, you are not obligated to follow this scheme; two such batteries are expensive, and the system will operate perfectly with just one of them. For that matter, the reader's personal pet robot construction may permit the use of a single, tall 24-amp-hour battery. It is a matter of taste and situation.

Figure 5-1 shows the interconnection of the batteries and related systems. The straight, unregulated 6 volts goes to the relay socket-strip, to supply the motors. The 6 volts is also regulated to provide a 5-volt, one-ampere supply, which in turn supports the brain board and the relay coils. The batteries are accessible by a one-quarter-inch phone jack mounted on the bottom or rear of the mainframe, to permit recharging processes. Lastly, a special voltage monitor circuit on the brain board keeps track of the battery status, informing the robot pet of critically low output, so that the pet may in turn inform its owner of incipient "hunger."

THE VOLTAGE REGULATOR

It's time to discuss the nature of the voltage regulator in fuller depth. You will remember that the average regulator has internal inefficiencies which make a 6-to-5-volt regulation impossible. This is due essentially to the nature of the transistor as a device, and the way it is utilized in a standard regulator. A transistor is a variable resistor of sorts, which, by the application of a proper bias current, becomes equivalent to a diode in terms of current-carrying capacity. In other words, depending on the bias current, a transistor moves from open-circuit to partial conduction and finally to conduction as if through a diode. This is how a regulator can control voltage flow.

But a diode is not a perfect conductor; it usually introduces a loss of about 0.7 volts. Likewise, a transistor in a regulator will conduct more-and-more up to a limit, a limit involving a 0.7-volt loss. This loss is compounded by the fact that the main power transistor in a standard regulator is *controlled* by *another* transistor with similar losses. This effect snowballs until, as we've seen, a +5-volt regulator loses about 3 volts just while operating.

There is, though, a way to get around the *saturation level*, the voltage loss, of the transistor. Ordinarily, the losses of regulator transistors add up, because the losses of each one prohibit the next one from operating at maximum level. But if the main power transistor could be biased fully on *normally*, with other transistors only there to turn it *off*, the losses would in effect cancel. That is precisely how voltage regulation is handled in the robot.

Figure 5-2 shows the complete schematic for the +5-volt regulator for the pet robot system. The main *series pass element*, that is, the main transistor which resistively drops the input voltage to a desired level, is a PNP power transistor in a TO-3 case, equivalent to 2N456B. You will notice that it is ordinarily biased fully on by the 68-ohm resistor to ground. (This actually "overbiases" it a bit, since ground is effectively lower than the collector; the end result is that the saturation level of the transistor is only two-tenths of a volt.) The op amp senses the output, and attempts to compensate for any deviation from an output of +5 volts. The 2.8-volt zener diode provides an accurate reference voltage for the comparison.

A smaller PNP transistor, equivalent to 2N3906, is linked between the input +6 volts and the power transistor base. As the 2N3906 begins to conduct (as controlled by the op amp), it gradually pulls the power transistor base away from ground, shutting it off.

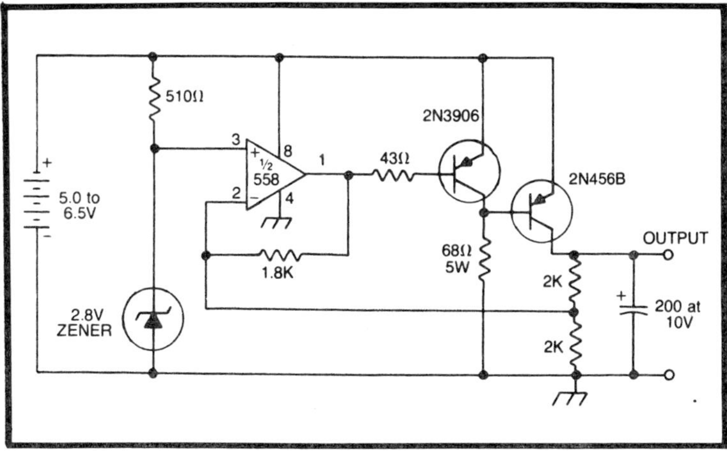

Fig. 5-2. Minimal input regulator for the power system.

Thus, if the op amp senses too great an output from the circuit, it can use the 2N3906 to reduce the output. Remember that individual differences in the transistors chosen will affect the stability of the circuit. After all, transistor gain plays heavily into the functioning of this approach to regulation. Therefore, the resistor values may require some experimentation during final test, particularly in the case of the output divider and the 5-W bias resistor.

The only worrisome thing about the circuit is that it is not failsafe. That is, if the 2N3906 or the op amp should ever fail, the natural tendency of the circuit is to turn fully *on*, not off...and though standard TTL circuits could withstand the over-voltage of 6 volts, the Intel 8085A and its peripherals most likely could not. So you may wish to insert some sort of over-voltage protection between the regulator output and the dependent circuit. The best approach would no doubt be to place an SCR across the output, wired so that an over-voltage condition would trigger it. The resulting short would blow an in-line fuse placed between the regulator and the circuitry. This standard "crowbar" method would result in temporary shutdown of the pet, but no serious (and expensive) damage would result.

Regulator Construction

Figure 5-3 shows the general construction of the prototype version of the regulator circuit. The main series pass transistor is mounted on a piece of aluminum angle, but it is electrically isolated from it, using a mica washer and other such hardware, since the transistor case is tied to the collector, which must *not* be grounded.

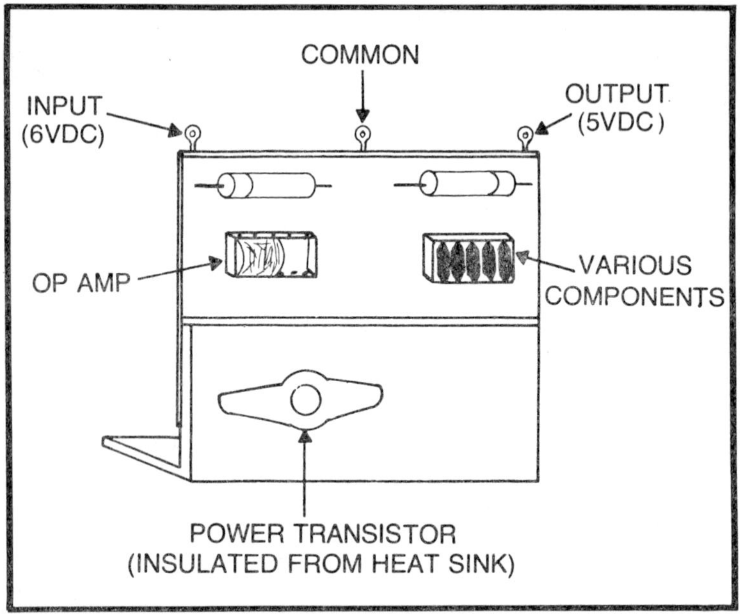

Fig. 5-3. Physical construction of the minimal input regulator.

In the absence of such insulating hardware, it is possible simply to mount the transistor to the angle, but electrically isolate the angle from the pet's grounded mainframe. The angle acts as a heat sink for the transistor, though ordinarily the transistor will not get too hot. After all, at worst it will be "burning off" only about one volt at an ampere, which amounts to one watt, enough to be mildly warm. The circuit board mounts to the angle using the same screws as the main transistor. The circuit board holds two 16-pin DIP sockets, one for the op amp, the other for minor parts. The 2N3906 and the large 68-ohm bias resistor mount directly to the main transistor at the appropriate terminal pins, and so are not visible in the illustration.

The capacitors chosen aid in the stability of the circuit in terms of ripple reduction and prevention of op amp feedback oscillation. Some experimentation with your own circuit may necessitate some variation in value, but the parts as chosen in this chapter are at least a good starting optimum.

The regulator circuit can be mounted in a number of places, but the most logical location is directly behind the batteries in the mainframe, which is a fairly empty area to begin with. Figure 5-4 depicts this choice of location in the prototype version. By using a standard fuse post, the in-line fuse, if included, can conceivably be mounted so that it is externally accessible. Yet the fuse should blow only under quite irregular circumstances. Use 18-gauge wire or so for connections between the batteries and the regulator to minimize line losses. Connections from the regulator output to the mounted connectors of the brain board should be as heavy a wire as is practical, though the terminals of a DB-25 connector will really only readily accept a 22-gauge wire. Wiring to the relay coils from the

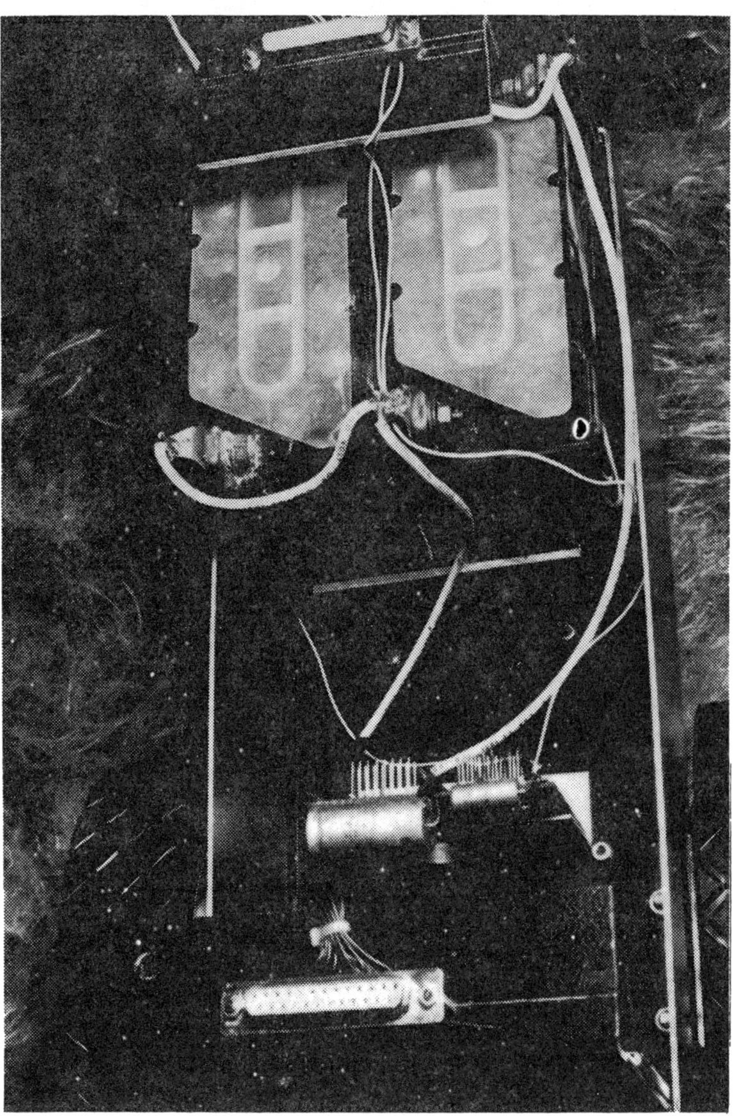

Fig. 5-4. Location of the regulator behind the batteries within the mainframe of the pet robot.

regulator may be light gauge, since those lines will carry a maximum of only about 20 milliamps. The connections from the batteries to the relays, on the other hand, may have to carry a few *amps*; make them heavy, perhaps 16- or 18-gauge. Finally, there needs to be a connection from the batteries to the brain board connector, since the battery-level sense circuit will be built later onto the brain board; any gauge may be used, such as 26-gauge, since the power required is negligible. Consult the pin configuration data in Table 4-1 of the previous chapter, when necessary.

A final connection to the batteries is the battery charger access socket. Place this socket somewhere in the mainframe of the robot pet, preferably near the DB-25 access connector mounted in Chapter 4. A standard ¼-inch audio phone jack is used for simplicity; the currents involved will not be too severe. The positive terminal of the jack is connected with 16- or 18-gauge wire to the positive terminals of the batteries. The jack itself is mounted and thus is in electrical contact with the mainframe, providing a pathway to complete the charging circuit.

CHECKING THE POWER SYSTEM

One final word on the subject of troubleshooting the power system as described herein: It is wise to put the regulator through its paces before installation. Build the prototype design and connect it to a 6-volt battery (at least) or a variable DC supply capable of simulating an output of from 5 to 6.3 volts (at best). The latter is a good procedure, since the batteries at full charge may be as high as 6.3 volts, and at lowest reliable output are about 5 volts.

Place a 5-ohm load on the regulator output, thus providing a test load condition of about 1-amp. Under such a load, the output should be no higher than 5.25 and no lower than 4.75 volts, through an input range of from 5 to 6.3 volts. This ±5-percent tolerance is absolutely necessary for the dependable operation of the RCU-85-related ICs. If the circuit acts erratically, check the output with an oscilloscope for oscillation and experiment with the values of capacitors and resistors chosen. If some sort of crowbar over-voltage protection will be added, test it out. It should be responsive and quick, but not so sensitive that minor fluctuations will cause continual, unwanted shutdowns.

The power system of the robot pet is now almost complete. It has all of the necessary voltages it may need, and some protection againt possible failure of the system. There is one all-important item now lacking, and that is the battery charger itself, the pet's "feeding trough."

Chapter 6
The Charging System

The robot pet now has a reliable power system, but it needs some way to replenish that system, to "feed," as it were, in further analogy to actual life. But it is not enough to provide just a battery charger, is it? After all, how will it know when it *needs* charging? And how will it get to the feeding trough when it's ready?

These are by no means empty questions. We have harped for many pages about the concept of life simulation, of the need to analogize to the way true living things operate. If we have any choice at all in the matter of charging systems, it would be consistent with our design philosophy to choose an approach that would simulate the way an animal would eat.

As I watch a pet dog in the throes of evening hunger, I notice a sequence of events:

1. Rover begs for food.
2. I prepare the food.
3. Rover walks over and eats the food.

Step One is simplicity itself as far as the pet robot is concerned. When the brain board is finally constructed, the voltage sense circuit mentioned in the previous chapter will be included. This circuit will alert the pet's RCU-85 computer to the fact that the batteries are low. The RCU-85 programming can then initiate an alarm sequence, some sort of peculiar "bark" which signals the fact of "hunger" to the human owner. It's no problem, then, to make our robot dog beg for food.

Step Two is no problem, either. Once the charger unit itself is constructed, the necessary charging voltage is available at the touch of a switch. Just pull out the powercable, whatever it may look like, and it's "dinnertime."

But Step Three is fraught with difficulty. How does the robot pet just walk over and plug in? We take for granted the maneuverability and senses of the average hound; he could find a bowl of dogfood in a howling gale. But the pet robot has limited perception and limited maneuverability. These assets are excellent as far as robots go, but crude and imprecise in a dozen ways. You can place a plug on the charger and a jack on the pet robot, but docking is so complicated, it's impractical.

It boils down to this: In the robot dog prototype, the pet can alert its owner to its need for charging, but the owner must bring it to the charger and plug it in. This is a definite trade off, a concession to simplicity at the expense of the design philosophy, but there it stands.

And yet, something of the sort of behavior described in Step Three has been accomplished by a number of clever roboticists, and far be it from me to discourage the reader from pursuing the idea further. (Eventually, I plan to spend more benchwork on it.) In fact, perhaps a bit of resource-pooling at this point might get us on the track of the problem. Let's try.

FEEDING-TIME APPROACHES

First of all, there's the matter of *point of reference*. The pet, when completed, will be able to avoid obstacles and such, but those obstacles are all relative to it. Putting it plainly, the pet robot has no way of knowing exactly where it is. It merely ambles about in a shadowy, unfamiliar world of ultrasonic echoes.

If our robot dog, though, *did* have some way of establishing its location, it could then follow an orderly, preprogrammed process intended to move it to the known location of the charger. Even if only the room could be established, it would help the pet to know how to get to the "meal."

There are no doubt a number of ways to approach the matter of point of reference, both in software and hardware. A software approach might be to have the pet robot memorize the approximate size of each room it might encounter, based on readings from its Soniscan ultrasonic system. Or it continuously could keep track of exactly how far it has gone, and in which directions, from the charger, so it could instantly find the shortest possible route along which to "backtrack." Both of these approaches have merit, but both require a good deal of software and memory storage, for which the reader must be willing to provide if he chooses to pursue either course. Plus, both concepts make certain assumptions which may or may not hold true. For instance, in the first approach it is assumed that no two rooms are of roughly the same size. In the second approach, it is assumed that the pet robot has some exacting way of determining the distance covered, such that little inaccuracies over the duration of 8 to 12 hours do not add up to total indirection.

There are hardware methods to establish a point of reference for the pet, but each has its own peculiar pluses and minuses. It is possible, for instance, to mount a pair of phototransistors on the underside of the mainframe, and then lay down a stripe which leads directly to the charger. When hungry, the robot dog would simply roll a bit until it would encounter the ubiquitous stripe, and follow it straight to "dinner." This is certainly the simplest way to handle the affair, but it has the non-ideal drawback of necessitating a major physical change in the decor to accommodate the pet. After all, if

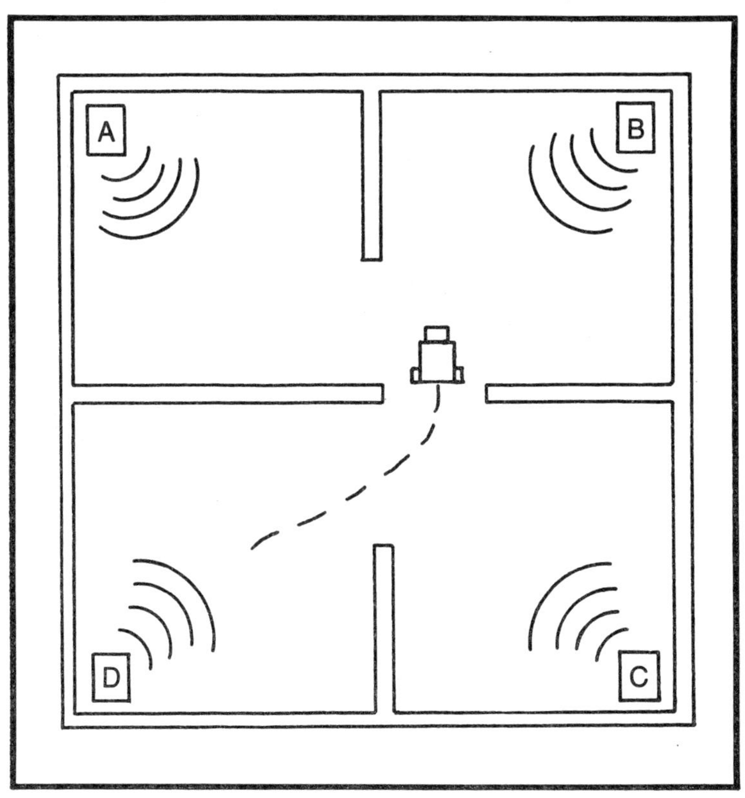

Fig. 6-1. Local-marker method of orienting the pet robot.

there's any other way to do it, one that does not require running an ugly stripe of tape through all the hallways of the home, we should opt for the other way.

A more interesting possibility is to place something in each room which specifically identifies that room, and then equip the robot pet to be able to detect the markers. For instance, each room might have a small box plugged into one of the wall outlets; the box would merely broadcast, either via ultrasonics or low-power radio signal, a marker frequency. Each room would have a different marker frequency, and the pet would then know where it was, depending on what it received at the moment. Figure 6-1 illustrates the concept. This is a farily reliable solution, but requires the design and construction of those little boxes (which, depending on the number of rooms, could be quite a task.) and the equipping of the robot pet to receive and differentiate the signals. Problems of *crosstalk*—when the pet receives more than one marker frequency at a time for one or another reason—would also have to be handled.

Perhaps a better angle from which to approach the issue, rather than from point of reference, is from the perspective of the charger itself. That is, it doesn't really matter where the pet robot is, as long as the charger can somehow tell the pet where *it* is.

Let's say, for instance, that the charger put out an ultransonic signal, and the pet robot had built-in equipment to home in on that signal. That would be half the battle, but then there's the problem of distance. The signal could only be "heard" so far, and the robot pet might well be in a faraway room when the "hunger pangs" strike. It might search and search, but never get on the track of the signal. On the other hand, the signal could be a medium-power radio signal, one that would reach through the very walls of the home. Then a new problem exists; it would head straight for the signal—and run straight into an intervening wall. There would actually be cases in which the only way to reach the charger would be to retreat from it for awhile, then advance on it again. Once again, the pet would get stuck in an eternal loop, knowing the absolute direction of the charger, but being unable to reach it.

One final complication of the subject of self-charging is the problem of actual hook-up. The pet's positioning skills are far from precise. The best it could do is head in the general direction of the charger, and perhaps hit it broadside. For it to bring about the precise mating of two small connectors is all but impossible. Some method would have to be devised by which the charger could manipulate the body of the pet robot into the proper position. A scheme like this could work, but would probably be very unreliable.

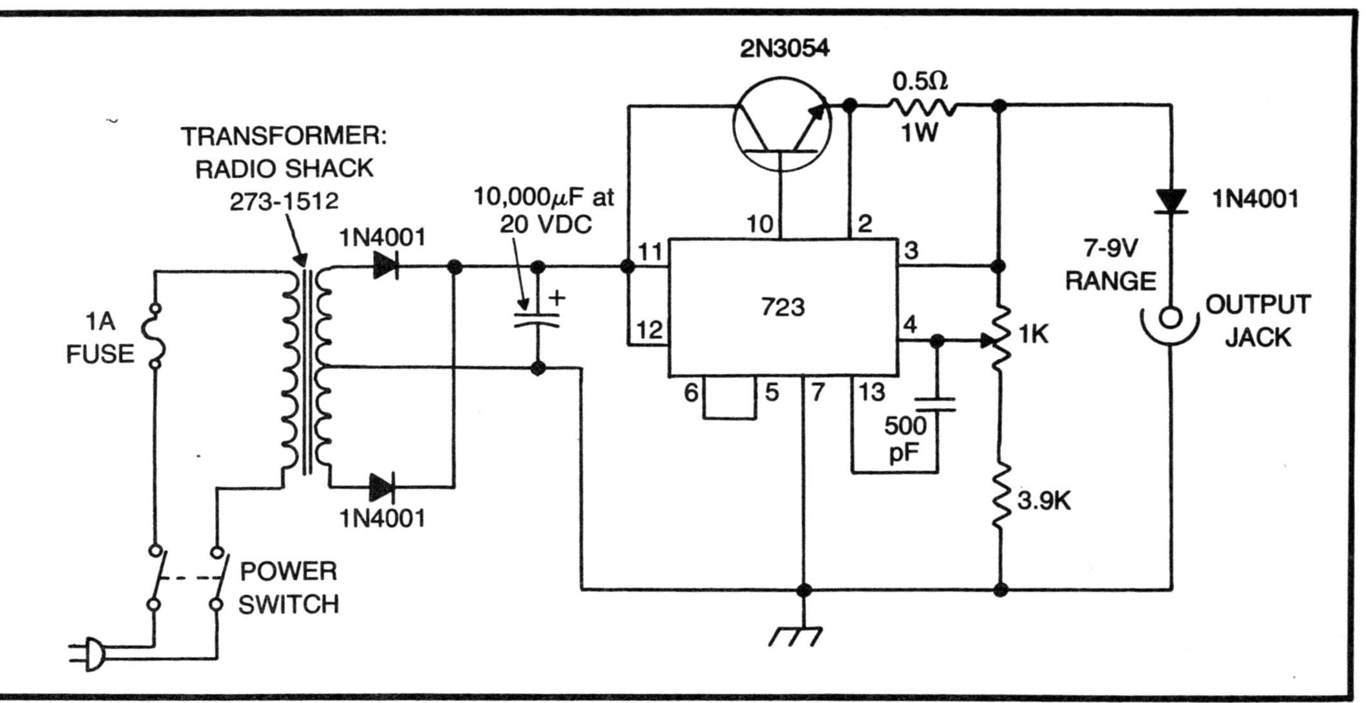

Fig. 6-2. Battery charger for the power system.

Again, let me stress that you're free to experiment and be creative. The difficulties I have cited are all improbabilities, not impossibilities. This is merely one more area in which the robot pet can be said to be a "research field." There is still much more progress to be made, and there is plenty of room for every interested would-be roboticist.

CHARGING SYSTEM

With this subject thoroughly discussed, let's take a look at the basic prototype battery charger system used for the pet robot. Figure 6-2 shows the schematic for it. The transformer and rectifier diodes provide a DC supply, which is regulated to about 7.0 volts by the 723 regulator as shown. The 723 is wired to prohibit the supply from sourcing any greater than about 1.2 amperes of current. This is a nominal limitation based on the variety of battery used; some batteries may be able to handle higher charging currents, some not as high, and the value of resistor R_s should be adjusted in inverse proportion according to the formula as follows:

$$R_s = \frac{0.65\text{V}}{I_L}$$

where R_s is in ohms and I_L is the desired maximum limited current in amperes.

A number of modifications are possible which make the charger more or less semi-automatic. Figure 6-3 shows two such additions. The first circuit would effectively be a charge timer, which would shut down the charger after a preset time limit of a few hours. The time limit would depend on individual experimentation, based on the charging characteristics of the specific batteries you choose.

The time duration follows the equation:

$$T = 1.1 R_A C,$$

R_A is in ohms, and C is in farads.
where T is in seconds,

A bit more of a deluxe version of charge-control is the second circuit shown. In this version, the circuit actually monitors charging current. When the current crosses some preset lower threshold, such as 100 milliampers, the state of the 311 comparator switches from high to low, thus releasing the relay and shutting down the charger. (Resistor values are subject to experimentation.) The 311 output could conceivably control the resetting of a 555 timer, as in the first control circuit, thus merging the two approaches. The

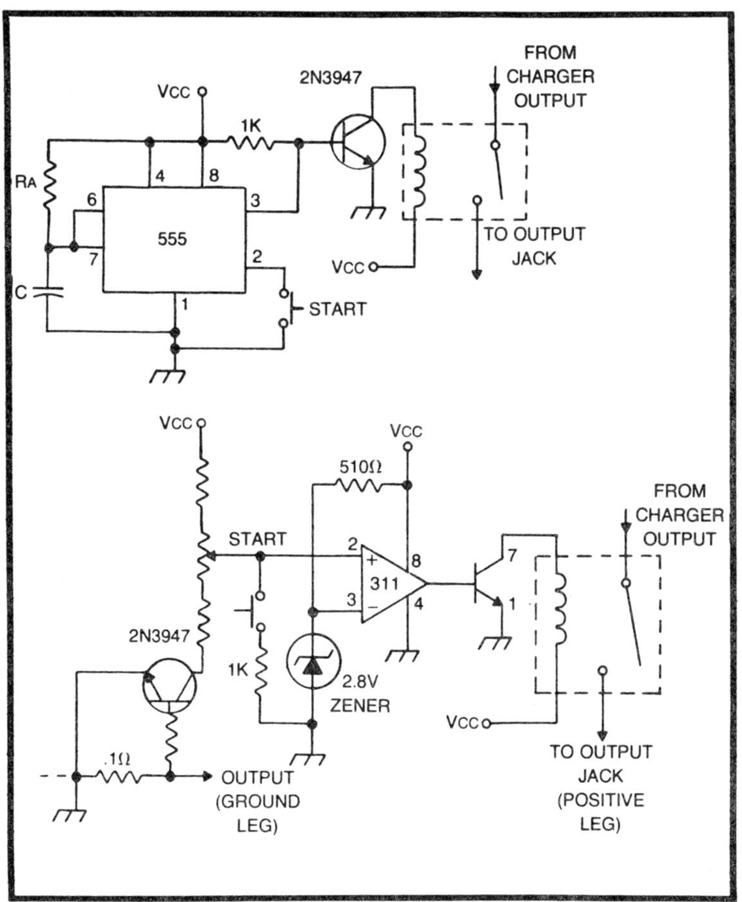

Fig. 6-3. Battery charger control circuits.

charger would shut down when the current dropped below 100 milliampers, or after four hours, whichever condition occurred first.

The charger output is wired to a standard ¼-inch phone jack from which an extension cord can be run to the robot's battery access jack. It is a good choice to use an extension cord with right angle phone plugs, since it otherwise might be difficult to plug it into the hard-to- reach access jack. Use a pair of 16-or 18 gauge insulated wires for the charging cable for the least possible resistance losses, and keep the cable short, no longer than a few feet.

While charging the pet, it's a good idea to reset the brain board (once it is incorporated), using the external reset switch that will be added later. It would definitely be problematic if the robot pet tried to roll off while plugged into its charger! Of course, the software can be written such that, while charging, the RCU-85 computer is placed into a loop, only to be escaped by a reset. But it's best to be safe.

With the addition of the robot pet's charger, we have completed the whole of the pet's physical subsystems. It is now capable of motion, independent from external supplies, except for occasional mealtimes. What stands on the workbench is, in short, a body waiting for a brain, our next subject.

Chapter 7
The RCU-85: A Brain For The Body

The construction of the pet's motor-driven frame has not been without its rewards. There before your eyes stands a miniature golf cart capable of moving about in your living room, turning as you direct it. But, alas, it is not yet a robot. For a robot is, remember, a "life-simulator." None of us walks around under the direct control of someone else's leash. We have our own minds, our own senses, our own sets of goals and priorities. The robot pet will not truly simulate life without its own version of these characteristics. It needs, in short, a brain.

Without waxing philosophical, a brain is that which receives sensory input, interprets it according to some world-model, and responds via motor output on the basis of a system of goals. The key to all of this is *versatility*. The robot pet's brain, like any brain, must be versatile enough to handle several inputs, consider them and respond with a variety of outputs. The circuitry to do this adequately must be correspondingly complex. A few, even many, TTL ICs will not do—too inflexible. For true variety of response, the brain must be microprocessor-based.

There are many inherent advantages to a microprocessor-based approach. One, certainly, is *centralization*. Consider the analogy of a living animal, say a dog. A dog does not have separate brains and nervous systems to respond to pain, pleasure, hunger, and so on. Sensory input, muscular output, body-temperature regulation, and internal rhythms are all generated from a single brain. Certain parts of the brain may have different responsibilities, but those parts are not distinct, self-sufficient controllers. All are interrelated.

So it is with the microprocessor-based robot. The processor has its own clock or frequency, with which all of its activities are synchronized. All other subsystems within the larger brain can derive their timing from the central clock. Also, with all sensory feedback and output control being processed from a single location, logical sequences of action are much easier to generate; all system factors are available for consideration.

The versatility of the microprocessor-based robot lies in its *programmability*. Standard non-processor robots generally are limited in number or response-options. If it sees a light, it tracks it; if it hits a wall, it retreats; and so forth. The beauty of a programmable robot is that the builder can shape its response patterns in a highly-complex fashion. If a "micro-dog" sees a light, it can track it, avoid it, bark at it, or ignore it. If it hits a wall, it can back off, sound an alarm, hit it again, or shut down. Its reactions are as endless as the number of outputs open to it. And its entire set of priorities and choices can be totally restructured in a matter of minutes—by reprogramming. The processor design opens the door to a more effective simulation of life responses.

The electronics market *abounds* with microprocessors and peripheral chips, and I won't attempt any sort of comparison. We do, however, have a number of parameters to limit our choice. First, the microprocessor must require a single-supply, +5 volts, for our battery supply is thus designed. Second, it ought to be an 8-bit microprocessor. Four-bit units such as the TMS 1000 are a bit limited, and 16-bit chips like the LSI-11 are too complex for the magnitude of this project. Third, it ought to be fairly simple and self-contained in design. It must not require too many extra parts to be complete. Finally, the cost must be reasonable.

A number of chips will fill the bill, but the Intel 8085A was chosen for the prototype robot pet. The reason was that the 8085A has all of the advantages of the industry-standard 8080A, but is +5-volt-supplied, complete in one chip, and has a number of added interrupt capabilities. The reader may wish to design around, say the Zilog Z-80 or the Motorola 6800, but the 8085A was found to be more than equal to the task.

PROCESSOR INTERACTION

In any processor-oriented system there are three major interacting blocks: the processor itself, memory, and input/output ports (see Fig. 7-1). The *memory* stores the program for the system,

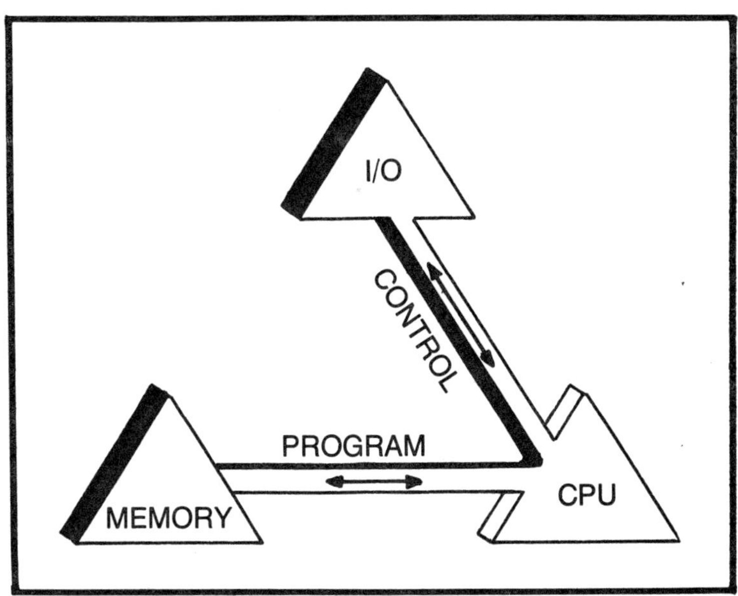

Fig. 7-1. The three interactive blocks of a processor-oriented system.

plus any data required for operation. The *processor* interprets the binary-coded program in memory and performs it, in terms of arithmetic functions, logical decision-making, and shifting of data. The *I/O ports* connect the system to the real world, permitting the processor to receive data and control external devices.

The two most common forms of semiconductor memory are Read Only Memory (ROM),and Random Access Memory (RAM). ROM is available to hobbyists in Erasable Programmable form (EPROM). EPROM can be programmed by a complex electrical process and erased by exposure to Ultraviolet light. Also, EPROM is "non-volatile"—that is, it retains its programming whether or not it is powered-up. RAM, in contrast, is "volatile"—it goes blank without power. However, RAM is easily programmed and erased via simple TTL levels from the processor.

From the hobbyist's standpoint, EPROM poses a problem. A standard EPROM programmer is an expensive investment, and most commercially-available programmers are designed for use with a personal computer system (which, if you don't own, you're out of luck). Plus, there is the matter of an ultraviolet EPROM eraser, which is an added investment. Some EPROMs can be programmed by a simple TTL circuit, but they are not very inexpensive. For these reasons, it was deemed desirable to attempt an all-RAM system for the robot pet.

But what about the "volatility" of RAM? Who wants a machine that has to be reprogrammed every time it is turned on? The answer is that, ideally, the robot pet will *always* be running. With its battery system, it can run for a good 12 hours without a charge. During night hours it can be hooked to its charger. Plus, as we'll see, part of the pet's "feeding station" will be a cassette player with an interface to the robot pet, enabling its memory to be restored within a couple of minutes. This may seem a nuisance, but it is practical, and certainly less expensive than the EPROM route.

Greatly simplifying things is the fact that the Intel 8085A "family" includes a chip termed the 8155. This versatile IC packs the following goodies into a 40-pin package: 256 bytes (words) of RAM; two 8-bit I/O ports and one 6-bit I/O port; and a 14-bit, fully programmable binary timer/counter. The robot pet will use two such chips, with a socket for a third if expansion is planned.

Home computing enthusiasts will obviously be scoffing at this point. Three 8155 chips will only provide 768 bytes of memory. "My home computer requires 4 *kilobytes* of memory just to play Lunar Lander! How can you control a robot with less than a K (a thousand bytes) of RAM?" Consider first that home computers require a good chunk of memory to interface to a keyboard input. Any decent version of BASIC is bound to need plenty of room. Also, some sort of CRT interface is probably required as an output, which implies a need for *display memory*. The robot pet, however, does not communicate with the owner except by certain audible codes. Consider second that the home computer is *information-oriented*; it is a thinker, a mathematician, a data handler. The robot pet's brain, instead, is *control-oriented*, it is a manipulator, a prioritizer, a servo director. The robot pet needs to store no programs for floating-point arithmetic or payroll calculation; it simply interrelates a limited number of inputs to a limited number of outputs according to pre-programmed priorities.

THE RCU-85

Figure 7-2 gives the basic pin configuration for both the 8085A central processor IC and the 8155 multipurpose IC. Naturally, it is not possible to give a full set of specifications for the chips in this volume, but such information is available from the manufacturer. Also understand that the robot pet project does not fully utilize the vast capabilities of the 8085A or its peripheral chips. It is a bit like using a nuclear pile to boil an egg, but this needn't worry us. Better to use an "overqualified" processor which affords room to grow,

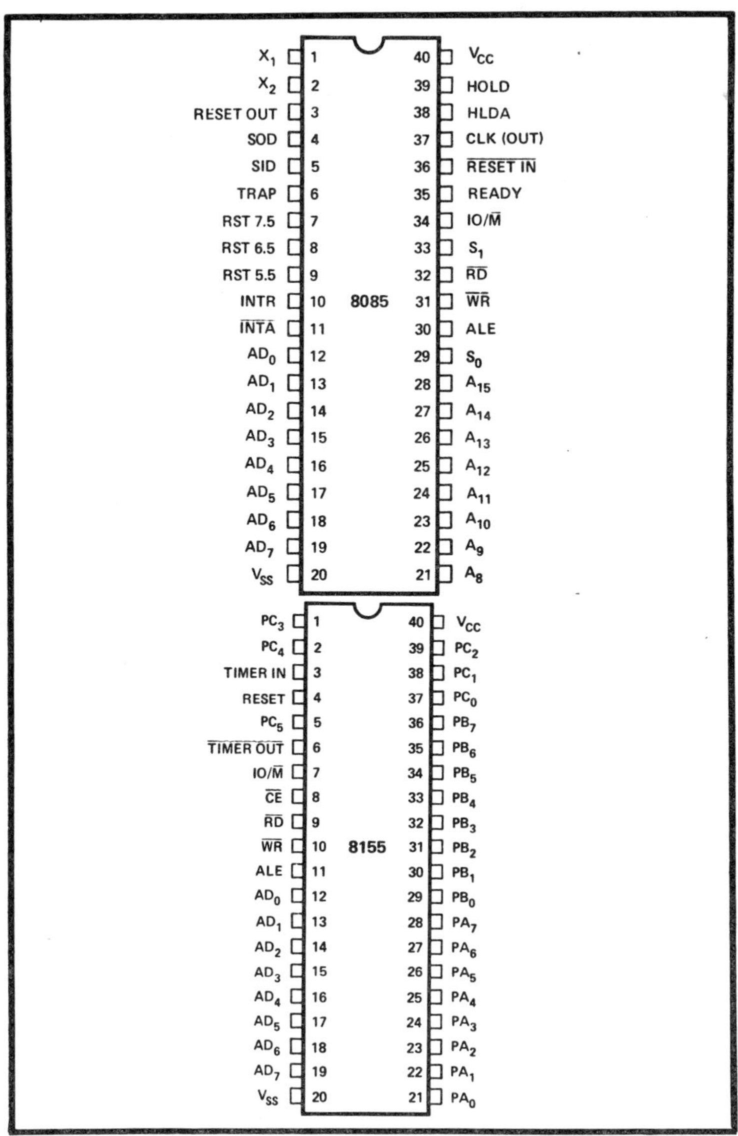

Fig. 7-2. Pin configurations for the 8085A and the 8155 (reprinted from the MCS-85 User's Manual, copyright 1978, courtesy of Intel Corporation).

than to use an adequate processor which might eventually threaten the project with obsolescence.

Figure 7-3 is the all-important schematic for what is termed the *robot controller unit,* or RCU-85 for short. It would've been somewhat confusing to try to illustrate the many, many lines which

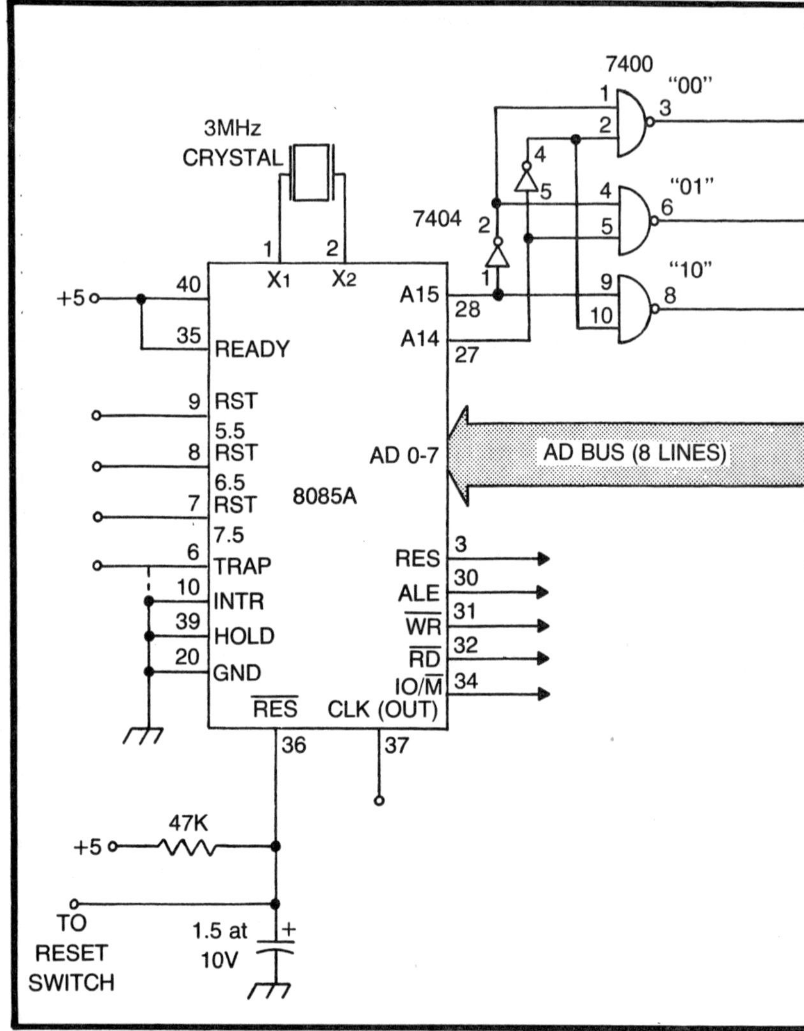

Fig. 7-3. The RCU-85, the "brain" for the pet robot.

interconnect the chips of the RCU-85. And so, we have illustrated certain features, such as the AD bus and the various ports, as arrows, even though each arrow in reality represents eight parallel wires. Such pin numbers as have not been noted in Fig. 7-3 can be derived from the previous figure.

The RCU-85 consists of a crystal-clocked 8085A, up to three 8155 peripheral chips, and some minor TTL gating. The 8085A divides down the 3 MHz frequency at the X_1 and X_2 inputs by a factor of two. Thus, the processor has a clock frequency of 1.5 MHz, which

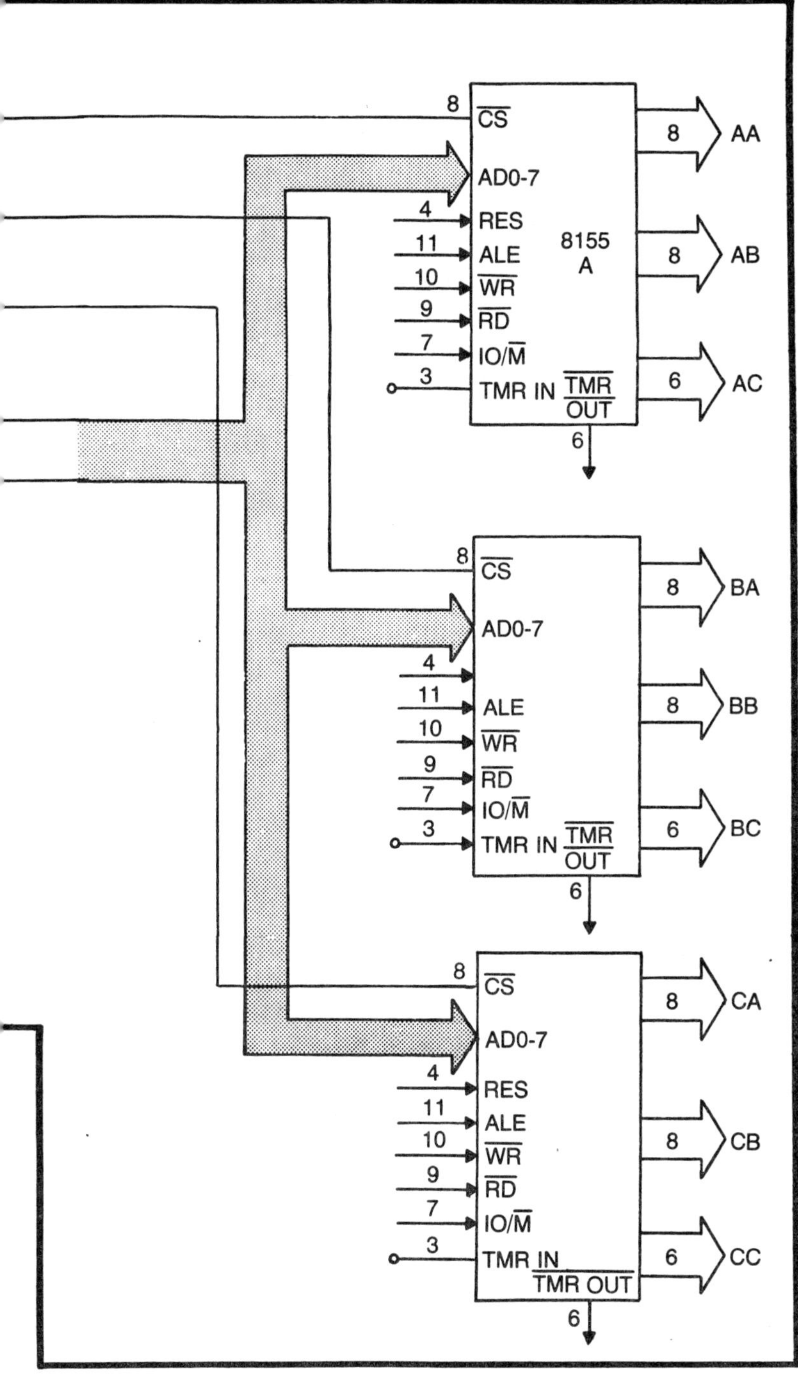

8155
A
CS
AD0-7
RES
ALE
WR
RD
IO/M
TMR IN
TMR OUT
AA
AB
AC
BA
BB
BC
CA
CB
CC

is available for external use at the CLK (OUT) output, pin 37. This is not the fastest that the 8085A will go; but for the minor decisions made by the RCU-85 in a real-time framework, it is instantaneous. Plus, as we'll see in Chapter 9, this particular frequency divides down well for application to the Soniscan circuitry.

The AD lines are common to all of the chips in the RCU-85. These convey 8-bit chunks of data to and from the processor, the ports, and memory. In addition, the processor uses these lines to send out an 8-bit *lower address* to signify which port or memory location it wishes to deal with. (This lower address is part of a complete 16-bit address; lines A_8 through A_{15} make up the *upper address*, which will be mentioned later on.) How does the system know whether the information on the AD lines is address or data? There is a common pin labeled ALE, for *address latch enable*, on all chips. The processor pulses this pin to signify that the information on the AD lines should be latched as address information. And so, the AD lines are time-shared or multiplexed, with the ALE signal to differentiate between the two types of information.

The 8085A has a few other pins which control the handling of information in the RCU-85. The $\overline{\text{WR}}$ pin (for write) drops to zero volts when the processor wishes to write an 8-bit word into a memory location or out of a port. Conversely, the $\overline{\text{RD}}$ pin (for read) goes to zero when the processor wishes to read from memory or a port. How, though, does the 8155 chip know which function is being read or written: memory or I/O ports? The IO/$\overline{\text{M}}$ pin solves this by going to a +5-volt level when ports are being dealt with, and to zero when memory is being handled.

Pin 36 of the 8085A is labeled $\overline{\text{RES}}$, for reset. This pin is normally held at a +5-volt level by the simple resistor-capacitor network shown in the figure. As long as this is the case, the processor will be running, executing programs, and handling data. But when this pin is grounded, the processor halts, the address counter built into the 8085A returns to zero, and data is no longer exchanged. (The robot pet will have a Reset switch mounted on its underside, near the recharge jack and connector "A" which have already been mounted.) Pin 3 on the 8085A is an output pin marked RES, which produces the very inverse of the $\overline{\text{RES}}$ input. When the processor is reset, RES goes high, and resets the 8155 chips, each of which has its own RES input. Memory is not affected, of course, but the ports are reset to an "all-zeroes" output.

ADDRESSING

We mentioned earlier that the 8085A really uses a 16-bit address, being conveyed by AD_0 through AD_7 and A_8 through A_{15}. The processor essentially has a 16-bit ripple counter, called the *program counter*, built internally. Under normal conditions, this counter begins at zero and counts upward, the result being output on the A lines and AD lines. In this manner, the processor begins reading through its memory sequentially, executing each binary-coded instruction it finds. The 8-bit lower address carried on the AD lines is common to all three 8155 chips; it accesses any one of 256 memory locations, but does not specify from *which* of the 8155 chips it wants to read. (Remember that there is a total of 768 words of memory with the three chips.) Here, then, is where the upper address on the A lines comes in. The highest two A lines, A_{15} and A_{14}, are decoded so as to select only one of the three 8155s at a time. 8155-A is read when the two bits are "00"; 8155-B is read when they are "01"; and 8155-C is read when they are "10". In this way, there are no ambiguities or conflicts, and no chance that any two of the chips will accidentally be enabled at the same time, which would result in a short. In program instructions that cause a "jump" to a memory location, a full 16-bit address is involved.

What about addressing I/O ports or the internal timers of the 8155s? The 8085A has two program instructions, IN and OUT, which are used to convey data to and from ports; in the 8155s, the timers are treated as if they were ports. In these two instructions, there is not a 16-bit address, as in memory interaction, but only an 8-bit word. This one 8-bit word, which the program specifies, is placed on both the A lines and the AD lines, to make up a *pseudo-16-bit* address to indicate specific ports and chips. Figure 7-4 shows how this information is used by the RCU-85 to pinpoint the very port or timer desired by the processor.

The upper 8-bit word is used the same in both memory and port addressing: the upper two bits are decoded to select one of the three 8155 ICs. In I/O addressing, though, the lower 8-bit word on the AD lines (which is identical to the upper word) is treated differently by the 8155s. When this lower word is latched into the chips by the ALE signal, the chips internally decode the lowest three bits of the eight, and use them to specify which port or timer is to be addressed. Port A has the code "001"; Port B is "010"; and Port C is "011". In addition, there is a special internal register termed the *command/status register*, which is coded "000". This C/S register can be written into to predetermine which ports are to be input and which are output, and to control the timer. Finally there are two 8-bit registers, coded "100" and "101", which are used to load the timer and set up its mode. Using the coded three bits, the program can select any of these ports or registers. Note, by the way, that the

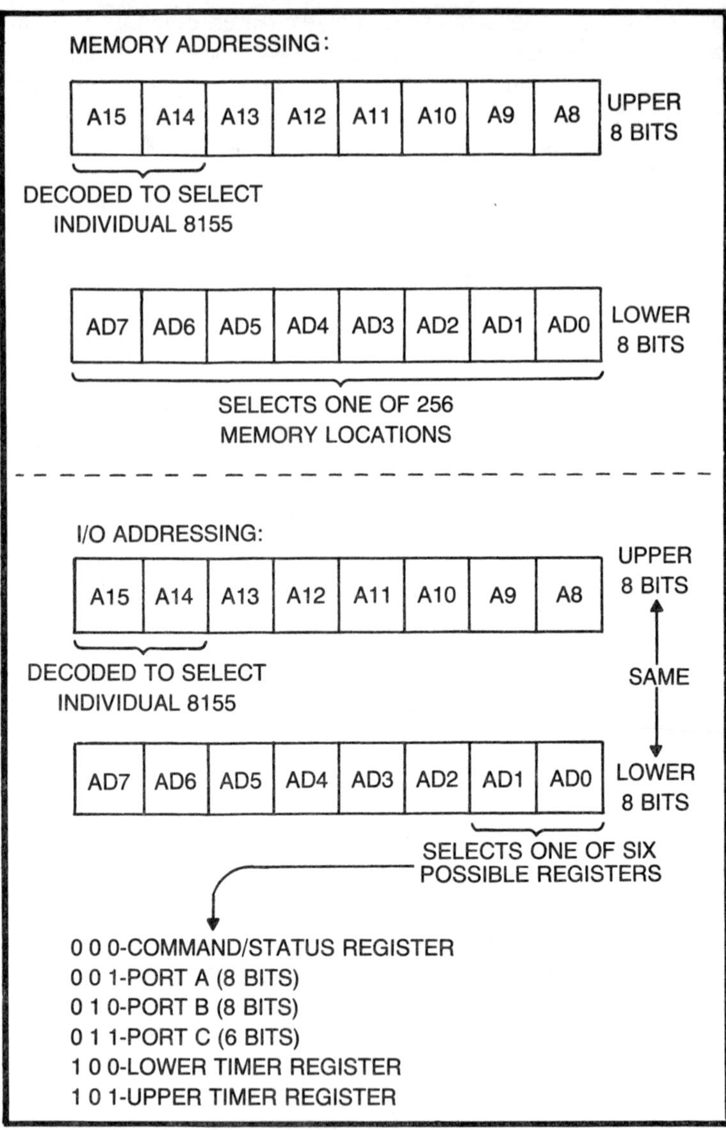

Fig. 7-4. Addressing of ports and memory in the RCU-85.

8155s ignore the other 5 bits latched on the AD lines as an I/O address.

Port Allocation

The previous explanation, though eventually necessary for proper programming, may have left you dizzy. Taking a step back, then, it is important to see what the 8085A can do. Under program control, it can retrieve 8-bit data words from memory or from input ports, and write 8-bit words into memory or output ports. All data is in the form of zero or +5-volt signals, basic TTL logic levels. The

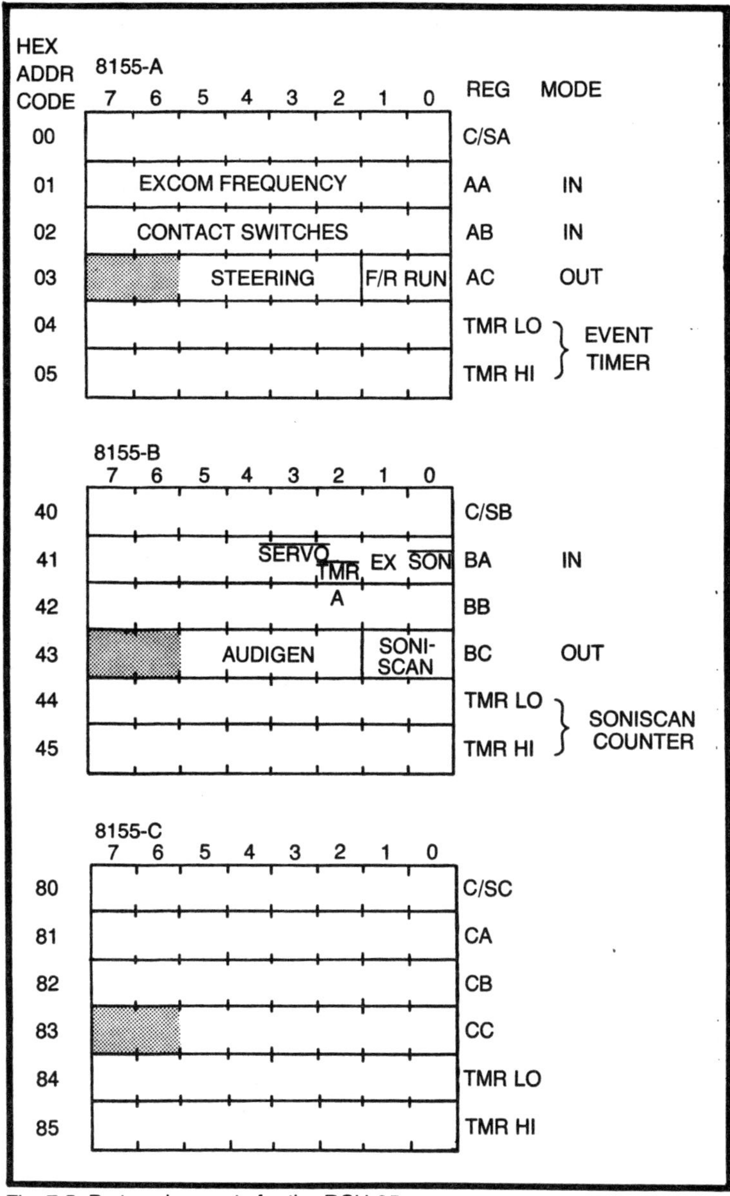

Fig. 7-5. Port assignments for the RCU-85.

brain board, then, consists of the central RCU-85, plus a number of TTL input and output circuits designed to link the processor to the robot's world.

Figure 7-5 illustrates the way in which the RCU-85 ports are linked to the peripheral circuitry of the brain board. This useful chart will be helpful both in programming and in system expansion. The center blocks represent each 8155, with the six internal registers, ports included. To the left is the I/O address of each register or port, coded in hexadecimal—units in first place, sixteens in second place. (Programming will use this sort of code, as a later chapter will explain.) To the right is the designation of each register, together with the mode in which it is functioning, input or output. Ports BB, CA, CB, and CC are not so designated; they are available for expansion. Notice, by the way, that Ports AC, BC, and CC are only six bits in size and use six pins on the 8155 package.

Several major TTL circuits complement the brain board and interface to the RCU-85 via these ports. The *Excom* circuit, for instance, has two outputs: an 8-bit word corresponding to a vocal input frequency, and a 1-bit "flag" indicating presence of a valid input frequency. The 8-bit output is wired directly to the eight pins that make up port AA; the 1-bit output is wired to the next lowest pin of port BA. The RCU-85 can use these incoming signals to enter the Excom mode. Further details on Excom appear in Chapter 10.

The Servodrive circuitry which controls the steering and drive motors of the robot pet has two inputs and one output: a 4-bit input to select a steering angle, a 2-bit input to control motion and direction, and a 1-bit output to signal the completion of a turn. The upper four bits of port AC convey programmed steering angles to the Servo drive circuit; the lower two bits of port AC activate a couple of the relevant relays; port pin BA_3 carries the completed-turn signal, termed $\overline{\text{SERVO}}$. (The Servodrive circuit is detailed in Chapter 8.) The upper four bits of Port BC fill the bill for the Audigen, or "bark" circuit, which requires a 4-bit input from the RCU-85. Chapter 11 handles the functioning of this circuit in depth.

The all-important Soniscan circuit, aside from its timer involvements, has an input and an output: a 2-bit input from the RCU-85 to select the direction of the ultrasonic beam, and a 1-bit output to the RCU-85 to signal the reception of an echo. The 2-bit control data is sent out of the lowest two bits of port BC, and the 1-bit "echo flag" is read from the lowest bit of Port BA. The Soniscan will be mentioned later, and is the subject of Chapter 9.

The remainder of the port pins are, so to speak, "up for grabs." Any conceivable input or output device which deals in TTL voltage levels can be hooked to these pins. If external contact or "bumper" switches are added to the robot pet's body for tactile sensing, these can be connected to the input pins of Port AB. If a wagging tail is added, the controlling relay or transistor can be hooked to a pin of an output port, such as port BB when it is properly programmed. The robot pet's body becomes a mobile "test bench" for any number of peripheral devices, each of which can be placed in communication with the centralized RCU-85.

Use of the Timers

The internal 14-bit timer/counters of the 8155 ICs are versatile in the extreme and quite straightforward to use. The 8155 has two pins related to the timer/counter: TMR IN, which clocks the timer, and $\overline{\text{TMR OUT}}$, which signals the status of the count. When the timer/counter is to be used, the program loads the two timer-related registers, TMR LO and TMR HI, according to Fig. 7-6. The lowest 15 bits represent the number down from which the timer is to count.

The highest two bits set one of four modes for the timer:

Mode 00—$\overline{\text{TMR OUT}}$ goes low during the second half of the count, and the count halts at zero.

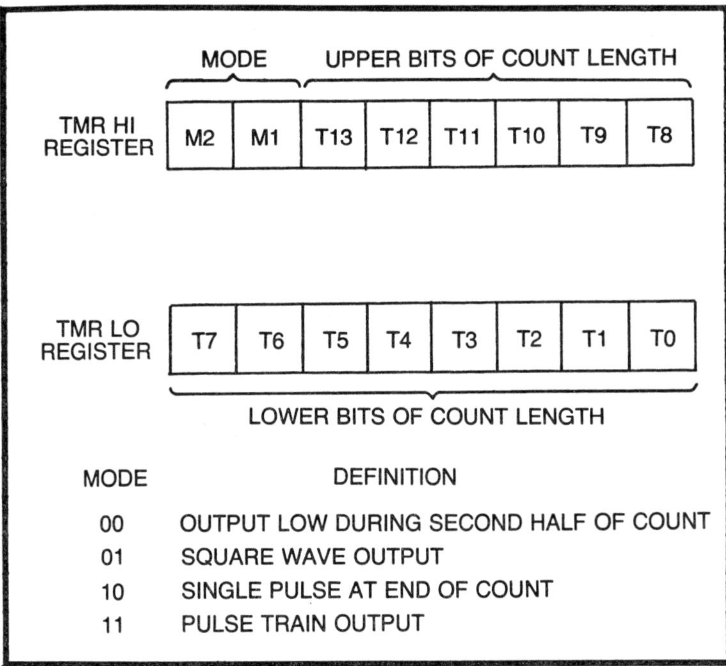

Fig. 7-6. Registers for the 14-bit timer/counter of the 8155.

Mode 01—$\overline{\text{TMR OUT}}$ goes low during the second half of the count, but the count continually retriggers at zero (square wave).

Mode 10—$\overline{\text{TMR OUT}}$ generates a negative pulse when zero is reached, and the count halts at zero.

Mode 11—$\overline{\text{TMR OUT}}$ generates a negative pulse when zero is reached, but the count continually retriggers at zero (pulse train).

The timer/counter is started and stopped by sending an 8-bit word to the C/S register after the proper numbers have been loaded into the TMR registers. The upper two bits of the C/S word are interpreted according to the description in Fig. 7-7. "00" does nothing; "01" stops the timer immediately; "10" stops the timer at the next reaching of zero; and "11" starts the timer, or restarts it after the next reaching of zero.

Once again, let us step back and see what this gains for us. If the pet robot wishes to delay a certain action, or wait for a given period of time, it has only to load a certain set of numbers into a timer and start it; then it can simply read $\overline{\text{TMR OUT}}$ at a suitable input port pin, such as BA_2 (see Fig. 7-5), to see if zero has yet been reached. This general *event timer* will be instituted in Chapter 12, and makes use of the timer/counter in 8155-A. Given an input clock at the TMR IN pin with a frequency of 10 Hz, the robot pet can negotiate time delays ranging from one-tenth of a second up to 27.3 minutes in increments of 100 milliseconds!

The Soniscan circuit is also simplified by the use of the timer/counter. As Chapter 9 will describe, the Soniscan divides the 8085A 1.5 MHz clock output down to a frequency usable for ultrasonic collision avoidance. This frequency must be divided down further by a factor of 8 bits (or 256), to provide a running 8-bit count of elapsed time from the instant of transmission to the reception of an echo. The timer/counter is just right for this task: the frequency is fed into TMR IN, and the timer is loaded with the number 256 (or, in 14 bits, 00000100000000). In Mode 01, the output at $\overline{\text{TMR OUT}}$ will be a square wave with a period of 256 times the input clock. When an echo is received, the RCU-85 can simply interrogate the number in the TMR LO register to see how far the countdown has progressed. Only the LO register need be checked, since the number will always be 256 or less, and the LO register covers this range of count. The timer/counter of 8155-B is used for this Soniscan function.

Other RCU-85 Inputs

The 8085A has one more crucial set of input pins, termed the *interrupt* section. The pins that concern us in the robot pet project are the four pins labeled RST 5.5, RST 6.5, RST 7.5 and TRAP. These pins, when raised to a +5-volt level, cause the program counter to jump to a specific address in memory, to perform a "sidetrack" program or subroutine. TRAP has the highest priority (should there be a conflict), and it jumps the program to address location 36 in memory. RST 7.5 is next in priority, and it jumps the

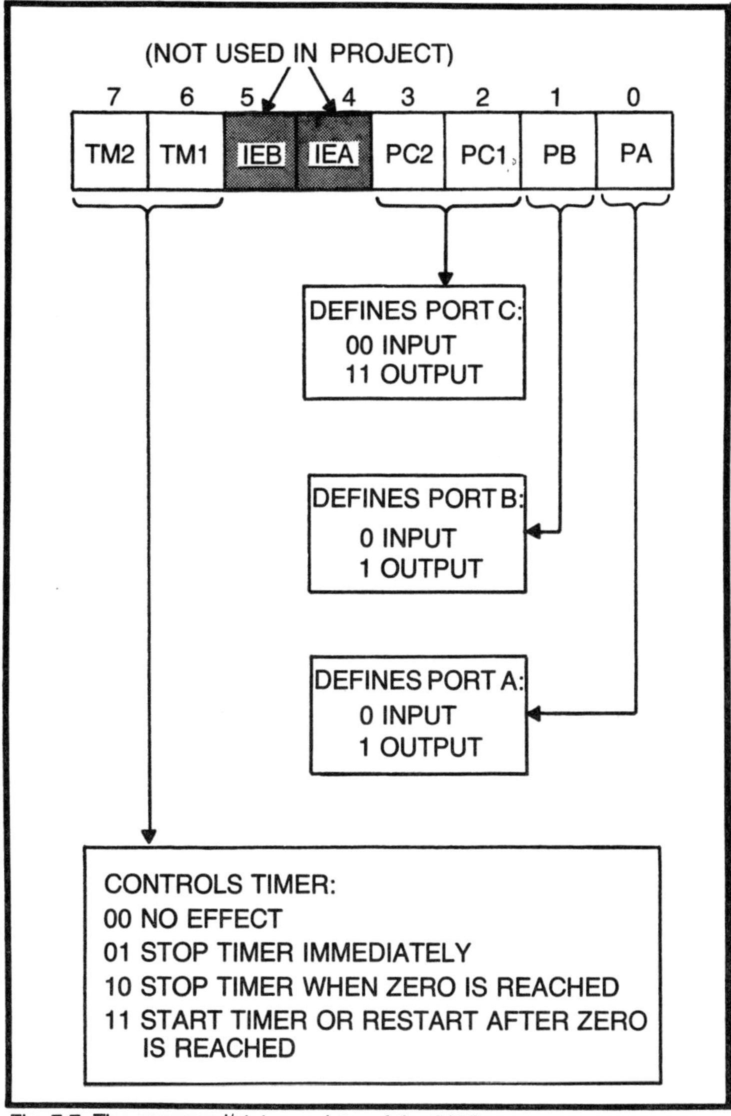

Fig. 7-7. The command/status register of the 8155.

program to location 60; RST 6.5 is next, jumping the program to location 52; and RST 5.5 is lowest, jumping the program to location 44.

The beauty of these pins is that the 8085A automatically keeps track of the place from which it was diverted, so that it may later return without missing an instruction. This is the reason for the term interrupt—side programs can be externally triggered without totally upsetting the normal program flow. The next address, the place to return to *after* the subroutine, is automatically stored in memory, in a stack of memory area set aside by the programmer. How this is handled is treated fully in Chapter 16.

Since TRAP is the highest priority, it is useful for "emergency programs, such as loss of power system voltage. The battery monitor circuit, which we will add in Chapter 12, uses this pin to sidetrack the pet robot into its hunger routine. Being a sensitive input, the TRAP pin should be grounded if it is not used. This is especially true since, of the four interrupt pins, only TRAP cannot be "disabled" by programming.

RST 5.5 is used by the Excom circuit to alert the robot pet to the presence of a vocal input. This pin is also jumpered to the previously-described port BA pin, for the following reason. Immediately after the interrupt is triggered, the program at location 44 "disables" RST 5.5, so that subsequent inputs will not keep jumping the program back to 44 *during* the subroutine. Meanwhile, the program reads the input informtion at the safer port BA pin. When the subroutine is finished, just before the program returns to its original course, the RST 5.5 is re-enabled and ready for a new Excom interruption.

The special *dual-mode* nature of the robot pet's activity, then, is based squarely on the interrupt system of the 8085A; without this sort of input, it would be nearly impossible to institute an effective vocal "override" concept.

The remaining two interrupt pins, RST 6.5 and 7.5, are open for expansion purposes, just as are any undesignated ports. There are a few other features to the 8085A which are not used in the robot pet project for one or another reason. INTR and INTA, for instance, belong to a basic interrupt system which is more complicated to use than the system already mentioned. SID and SOD are, in essence, 1-bit input and output ports, which are normally used in serial data communication applications—not applicable, at least presently, to the robot pet. HOLD and HLDA are used to synchronize interaction with slower-speed memory, but the 8155 RAM is fast enough for the system. Finally, S_0 and S_1 are status flags which indicate the

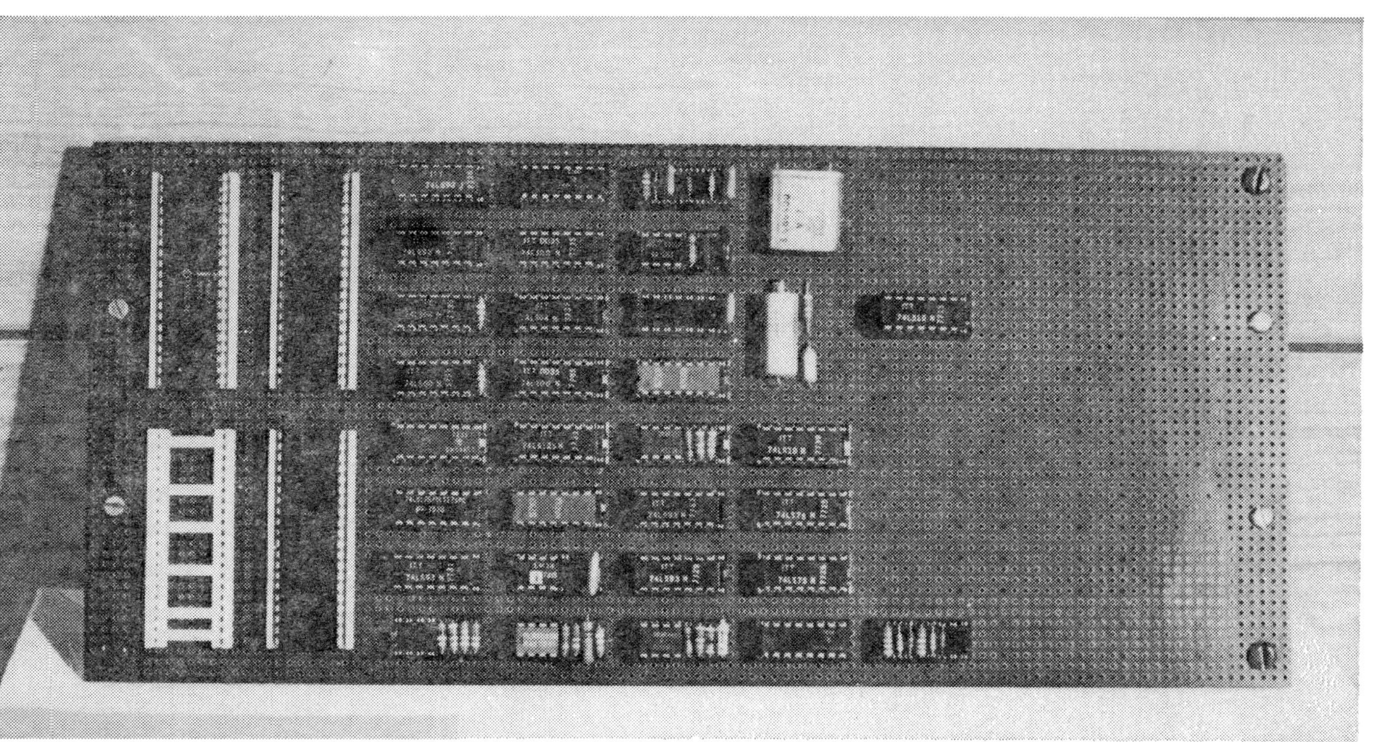

Fig. 7-8. The brain board.

type of data transfer presently occurring. The experimenter may be able to find unique uses for some of these features in an advanced pet robot version, and if so, he should contact the manufacturer and obtain a copy of the *MCS 85 User's Manual* for further details.

BRAIN BOARD ASSEMBLY

Figure 7-8 pictures the prototype brain board. A rectangular piece of perforated board, with holes on .100″ centers, was used; its dimensions are 5.00″ by 11.00″. To the extreme left of the board are the four 40-pin sockets which hold the 8085A and (up to) three 8155 peripheral chips. Near the upper center of the board are the crystal and reset network of the 8085A, plus a 22-mfd capacitor across the incoming +5-volt supply to reduce ripple current. (These centrally-mounted parts would probably better be mounted near to the 8085A, since they relate to it.) The remainder of the board is taken up with 16-pin DIP sockets to hold the TTL peripheral circuits which interface to the RCU-85 proper. Notice that there is plenty of room for expansion; the board as shown in the photo can hold some fifty or more DIP sockets comfortably. The single factor that contributes to this high-density situation is the use of wire-wrapping for interconnection. If the brain board were a double-sided printed circuit board, it would certainly require four or five times the space.

To the left of the RCU-85 section, below the board and thus not visible in the photo, is a DB-25P male connector. This is the connector which plugs into connector "C," which was mounted to the rear portion of the mainframe in Chapter 4. The major control signals of the RCU-85 are conveyed through this connector out to the access connector "A," mounted on the underside of the mainframe. The primary purpose is to provide access to the controlling signals of the RCU-85 for the purpose of external programming. As Chapter 13 will explain, the 8085A permits its control pins to go to an *open-collector* state when it is reset. Thus, during reset, an external device could simulate the control signals, place these onto the "floating" RCU-85 lines, and thereby program the internal RAM of the 8155 chips.

Likewise, to the extreme right of the brain board on the underside, there is another DB-25P, the mate to connector "I" which has been mounted toward the front of the mainframe. This connector acts as the go-between for any input or output of the peripheral TTL circuits on the board. Soniscan signals, Excom data, Servodrive feedback, Audigen output—all of these make use of this connector.

Each of the two connectors is mounted on .375″ standoffs, which bolt to an aluminum bar measuring 5.00″ by .375″ by .125″

thick. The aluminum bars add rigidity to the somewhat flimsy perforated board. If further strength is desired, two aluminum side rails may be mounted along the length of the board, uniting the sidebars. The only precaution in such a case is to leave room to prevent any shorting to the DIP socket pins. As noted in Chapter 4, there are only two hard-and-fast mounting rules for the connectors: they must be mounted 10.5 inches apart, so as to mate to the mainframe connectors, and they must be mounted in such a direction that the brain board could not possibly be plugged-in backwards.

The major task in assembling the brain board is the wiring of the RCU-85 section. Fortunately and conveniently enough, the designers of the 8085A family attempted to assign pins for the chips in such a way as to facilitate wiring. Notice that the power pins and the AD pins are the same for both the 8085A and the 8155. The $IO/\overline{M}$, $\overline{RD}$, $\overline{WR}$, and ALE pins for the two chips are at least in the same order on opposite sides of the chips. And so, it is a bit easier than otherwise to interconnect the 40-pin sockets. Much of the wiring between the sockets will consist of repetitive, parallel connections, requiring more patience than thought.

After the 40-pin sockets are mounted and wired, then the simple decoder for pins A_{15} and A_{14} can be added. Two TTL chips, a 74LS00 and a 74LS04, provide the necessary gates. You can mount and wire these two chips to the immediate right of the 40-pin sockets. The 8085A crystal and reset network, and the supply decoupling capacitor, can also be mounted near these chips.

One final addition is the reset switch, which is mounted beneath the robot pet near the charging jack and connector "A." Use a small SPST switch; tie one terminal to ground (the chassis) and tie the other terminal to pin 4 of the nearby connector "A." When the switch is closed, the line from connector "A" to connector "C" to the RCU-85 pulls the $\overline{RES}$ input of the 8085A to ground, resetting the processor. When the switch is open, the processor runs normally.

Much has been covered in this chapter, and much of it will not be fully appreciated until the robot pet is programmed. Yet, take the time to reread this chapter, and attempt to grasp the key concepts. The centralization of control offered by the microprocessor approach is the major factor that makes the sophisticated responses of the pet dog possible. If you should get discouraged somewhere in the midst of wire-wrapping the RCU-85, remind yourself that those four sockets, in and of themselves, will replace *boards* worth of circuitry. Once this realization dawns, it is time to move on to the peripheral support circuits, the first of which is described in the next chapter.

Chapter 8
Servodrive

Now that the rudiments of a microprocessor controller have been realized, in the RCU-85, it is time to link that controller to its "body." By using the manual controller link mentioned in Chapter 4, it has been possible to "walk" the pet robot around the living room. Now we can replace that controller with a circuit to do the driving and steering automatically.

Steering servo systems are always among the touchiest of robot circuits to implement. There are a few basic approaches, each with some maddening, built-in demon ready to confuse the matter. One approach is to have the steering column turn until a specified microswitch is contacted. This concept works well, except the steering motor may start up again if the motor overshoots the microswitch, unless your circuit is clever. Plus there is not much resolution in turning-angles—but, then again, how many turning-angles are necessary?

A second approach is a thoroughly linear servo system. This involves monitoring the steering-angle with a potentiometer linked to the column. By comparing this voltage to a control voltage by means of an op amp, a transistor drive circuit can be devised to position the column quite accurately. There are stability problems with this approach, but the major difficulty is in interfacing. The RCU-85 is "hopelessly" digital and doesn't mix well with analog voltages.

As usual, a hybrid approach provides the best workable alternative. The robot dog uses a voltage-feedback loop with a potentiometer, but handles the feedback comparisons *strictly digitally*, which reduces the instability and satisfies the digital tastes of the RCU-85. The key to the approach is the use of a very basic analog-to-digital converter. This unique addition to the robot pet system is termed the Servodrive circuit.

SERVODRIVE DESCRIPTION

Figure 8-1 shows the complete circuit. A potentiometer coupled to the steering column provides a variable reference voltage of from 1 to 4 volts. (You will no doubt recognize this as the servo potentiometer described in Chapter 4.) The output is lowest while the column is in its far left position, and increases as the column rotates to the right-most limit. This voltage is the feedback of the circuit as to the exact positioning of the drive wheel at any given moment. The feedback voltage is fed into one leg of a 311 voltage comparator IC.

Meanwhile, there is a free-running 7493 4-bit binary counter, being clocked by a reference frequency. The 4-bit output is fed to an inverter buffer stage, the 7416. The buffered output produces varying currents in a 4-resistor ladder, which causes the 558 op amp to produce varying voltages. The output of the op amp is a voltage "staircase" ramp, from about 1 to 4 volts in 16 increments. This ramp is fed into the other leg of the 311 voltage comparator.

The nature of the 311 comparator, as wired, is such that its output switches from 0 volts to 5 volts each time the 558 voltage ramp *passes* the potentiometer reference voltage. Thus, the output of the 311 is an asymmetrical square wave, with the positive-going edge occurring at the *exact moment* that the two voltages are the same. Figure 8-2 provides a comparison of the related waveforms.

It is then a simple matter to use a 74175 quad flip-flop to latch the 7493 counter output. Each time the ramp equals the feedback voltage, the positive edge of the 311 clocks the 74175 and latches the status of the 7493 counter. In this way, the 74175 output maintains a consistent *digital* record of the position of steering column. If the output were 0000, the column would be completely to the left; 1000 would be about middle; 1111 would be extreme right. We have, in fact, wired up a simple analog-to-digital converter.

The final piece in the puzzle is the 7485 magnitude comparator. This IC compares any two 4-bit words and sets its output pins according to the result. Using this IC, the 74175 "steering status word" can be compared to a 4-bit word written out of the RCU-85. If the two words are identical, all is fine; if not, the 7485 outputs can put things into motion to correct the disparity.

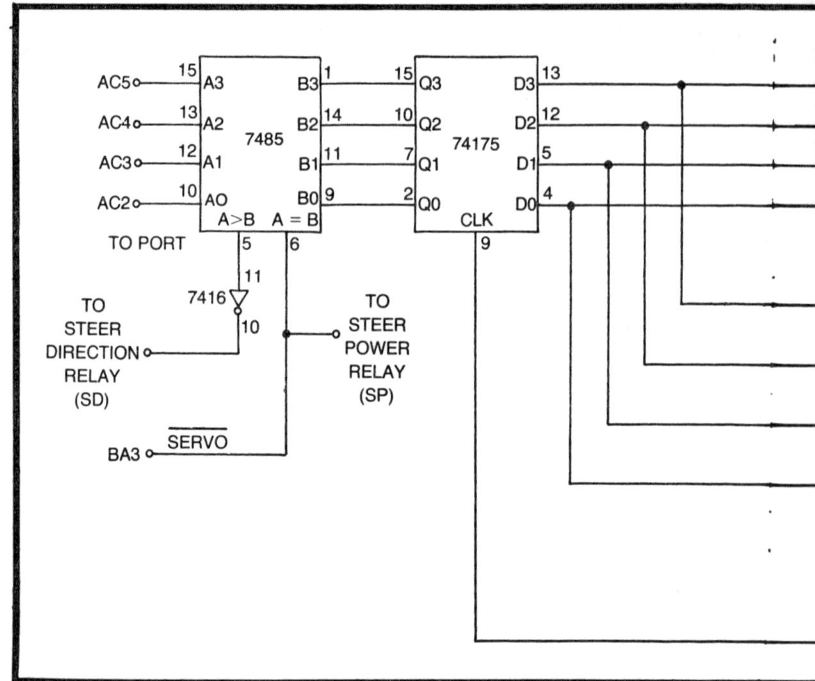

Fig. 8-1. Servodrive circuit.

There are three outputs to the 7485 comparator, two of which will be used herein. The "A<B" output goes to a +5-volt level if the digital word at the "A" pins is *less than* the binary number of the "B" pins. The "A>B" output goes to a +5-volt level if the comparison is just the opposite. And the "A=B" output goes to +5 volts if the two binary numbers are equal.

The various relays which control the steering of the pet robot require a zero-volt level to energize. And so, note that the "A=B" output of the 7485 controls the steer power relay (SP). As long as the steering column is positioned just as the RCU-85 has requested, the "A=B" pin will be at +5 volts, and the SP relay will be dormant. But if the RCU-85 issues a *new* binary word, one which disagrees with the present status of the steering column, the "A=B" pin will drop to zero volts due to the inequality...and the SP relay will energize the steering column to correct for the difference.

Now is when the "A>B" pin comes into play. Remember that the steering column position increases in numerical value from left to right, from 0000 to 1111. So if the new incoming word from the RCU-85 is higher than that of the column, the column needs to turn *right*. The "A<B" pin will be at +5 volts in this case, and the inverter

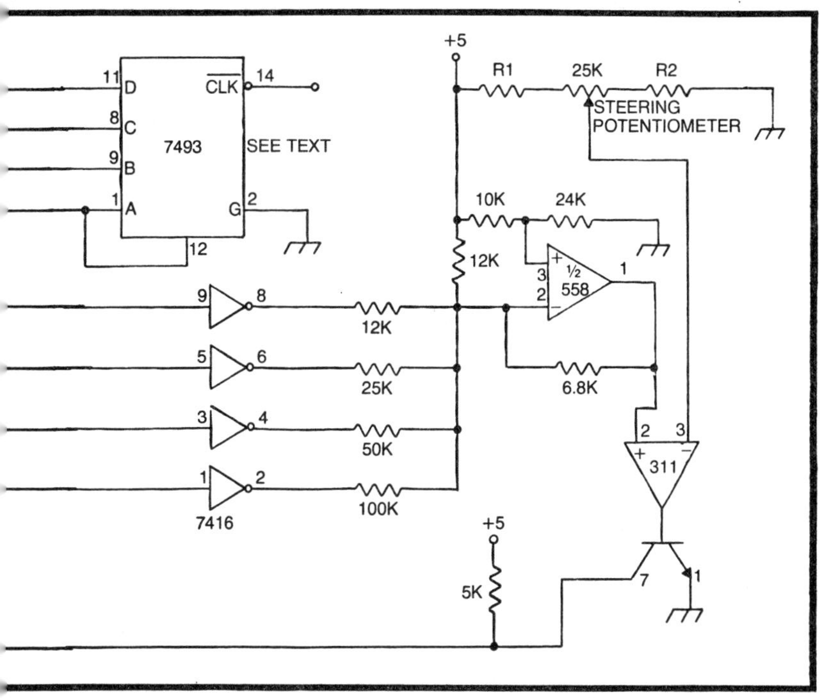

as shown will output a zero-volt level to the steer direction relay (SD). This energizes the relay, and the steering column is wired to turn *right* when the relay is so energized. And so the turn occurs, until an equality is re-established, at which point the "A=B" pin will go high and stop the column.

On the other hand, suppose the new number from the RCU-85 is less than the actual position value. In this case, the RCU-85 is requesting a turn to the left. The "A>B" pin is now at a zero-volt level, which is inverted to a high level. The SD relay remains unenergized, permitting a turn to the *left*, and a subsequent equality halts the column via the "A=B" pin. (Note that the "A=B" line is available to the RCU-85 via port pin BA3, so that it can know when the turn has been completed.)

At least two features contribute to the overall stability of the Servodrive circuit. The first is the fact that the motor, when de-energized, is immediately shorted by the opposite contacts of the SP relay; remember that this is how the relays were wired back in Chapter 4. This dampens the motor and prevents excessive overshoot which might cause the circuit, relays and all, to break into oscillation. Second is the resolution of the steering angle. The

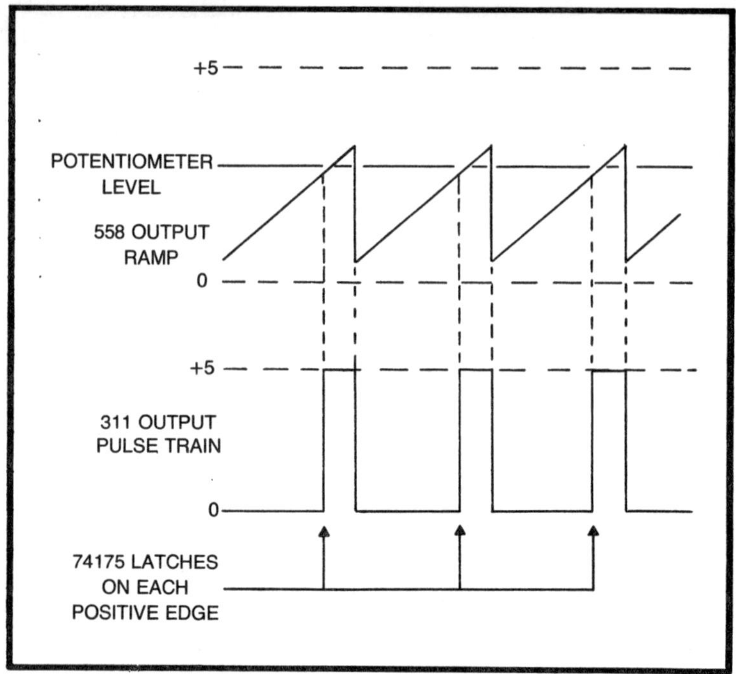

Fig. 8-2. Major waveforms of servodrive.

sixteen possible angles are more than enough for maneuvering the robot pet. And yet, it is coarse enough that a *bit* of overshoot wouldn't be enough to set off an oscillation in the feedback loop.

CIRCUIT WIRING

Wire-wrapping this reliable steer-control circuit onto the brain board is fairly straightforward. It requires about seven 16-pin DIP sockets: one each for the 7485, 74175, 7493, and 7416 chips; one each for the 311 and 558 chips with their related resistors; and one for the resistor ladder. Eighth-watt resistors can be used and easily formed to fit into the DIP sockets. The sockets can occupy any location on the brain board—but keeping them close together will aid in wiring and help to avoid errors.

The upper four bits of the 6-bit port AC have been designated for the steering function. Thus, pins 39, 1, 2, and 5 of 8155-A should be wired to pins 10, 12, 13, and 15 of the 7485 respectively. Since the 8155 ports are latched, the robot only needs to issue a 4-bit steering word once each time, updating it only when a change is desired. Also, connect pin 6 of the 7485 to port pin BA_3 of 8155-B, which is pin 24, to provide the RCU-85 with a means of knowing that the turn is complete.

There are a couple of matters to clear up before the Servodrive circuit can be completed. First, we said nothing of how the 7493 is clocked. This is because its clocking frequency is derived from a circuit not yet on the brain board, namely the Soniscan circuit. Figure 8-3, however, gives the pertinent parts needed to provide the clocking function.

The 8085A has an output, CLK(OUT), which provides a reference frequency that is one-half of the crystal frequency. Thus, if the 8085A uses a 3 MHz crystal, CLK(OUT) supplies a 1.5 MHz clock. For the purposes of the Soniscan, this frequency is divided down by a 7490 Divide-By-Ten chip and a 7493 Divide-By-Sixteen chip. The resultant frequency at pin 11 of the 7493 is about 9375 Hz. This waveform can then be used to clock the 7493 of the Servodrive circuit, providing a "staircase" ramp of about 586 Hz. The actual clock frequency used by the Servodrive circuit is not at all crucial, but it was found more reliable to run the circuit at this slower frequency, as opposed to using the CLK(OUT) frequency directly.

The second matter which requires attention is the choice of resistors R_1 and R_2 on either side of the servo potentiometer. Recall that the purpose of these two resistors is to adjust the output of the potentiometer to a sweep of about 1 to 4 volts, the same as the "staircase" ramp. The important factor here is to aim for an output range a bit *beyond* the ramp, or about 0.75 volts at full left to 4.25

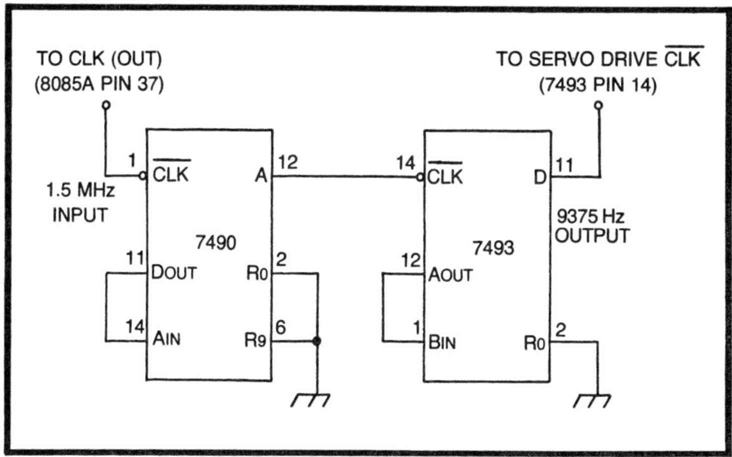

Fig. 8-3. Counters which provide a clock for the servodrive ramp.

volts at full right. This acts as a safety margin; if the range were tighter than the 1-to-4-volt ramp, then it would be possible for the RCU-85 to select extreme turning angles which would be beyond the potentiometer's rotating capabilities. That is, the RCU-85 might request an angle of 0000, which is about 1.00 volts; if the potentiometer could output no lower than, say 1.25 volts, the Servodrive circuit would never find an equality. The column would keep rotating until it jammed. And so, it is better to widen the potentiometer range a bit.

The best way to approach the matter is first to get the ramp "up and running," and then select the two adjustment resistors. Wire up the 7493, 7416, 558, and the associated resistors; wire up also the two counter ICs of the Soniscan circuit to provide the clock. (If the RCU-85 itself has not yet been wired, an external TTL square wave generator may be used.) Then monitor the output of the 558 op amp on an oscilloscope. Experiment with the values of the resistors labeled 6.8K, 10K, and 24K, to produce a clean, undistorted ramp from 1.00 volts to 4.00 volts. The 6.8K resistor controls the gain, or size, of the ramp; the other two resistors alter the vertical positioning or bias of the ramp. If a wide variety of substitute values are not available, you may wish to use small trimmer potentiometers instead of fixed values, to simplify the calibration process.

Once the ramp is fixed from 1 to 4 volts, it is a matter of some experimentation to choose R_1 and R_2 for the servo potentiometer. If a 25K potentiometer is used, and a range of from 0.75 to 4.25 volts is desired, $R_1 = R_2 =$ about 5.35K. If the potentiometer is not well centered, or if it is not exactly 25K, adjustments will have to be made empirically. The goal, though, is to view a slow, clean oscilloscope trace from 0.75 to 4.25 volts as the column is rotated from extreme left to right.

After these calibrations are made, we can attend to the rest of the circuit. Wire up the 7485 and 74175 as shown. Three lines lead to off-board locations: the two relay control leads from the 7485, and the center return lead from the servo potentiometer. These go to the brain board connector corresponding to connector "I," pins 6, 7, and 8, as specified by the connector pin chart in Chapter 4. One important notation—the wiring of the two relay control leads assumes that the special low-power relays are being used as described in Chapter 4. If you substitute a less efficient relay in your own design, you will certainly require some sort of driver transistor to buffer the "A=B" and "A>B" outputs.

When the entire circuit is wired up, and the power is turned on, you may be shocked to see the steering column immediately leap into a right turn. The reason is simple, when you think about it. The circuit merely obeys the output of the RCU-85. But the RCU-85 is as yet unprogrammed, and the four port-pins are in a high-impedance state, which the Servodrive circuit reads as 1111. Thus, it attempts to correct for this by making a right turn. The proof of the pudding, in this case, is if the column *stops* when it should. If not, quickly disconnect the power and check the wiring. If the servo potentiometer polarity is reversed, if the return lead is broken, or if the relays are malfunctioning, there will be faulty operation.

Perhaps the column does stop at the extreme right but then it jerks left and right in a slow oscillation, relays clicking. If so, try placing a small capacitor, perhaps .1-mfd, from ground to pin 7 of the 311 voltage comparator. This should kill any transients which may be causing indiscriminate latching of the 74175.

There is one more small circuit which forms a portion of the Servodrive system, and it is given in Fig. 8-4. This simple gating circuit links the drive power (DP) and drive direction (DD) relays to the lower two bits of the 6-bit port AC. Bit 0 energizes the DP relay and thereby the drive motor when it goes to a logical "1". Bit 1 controls the DD relay, permitting forward motion when at a logical

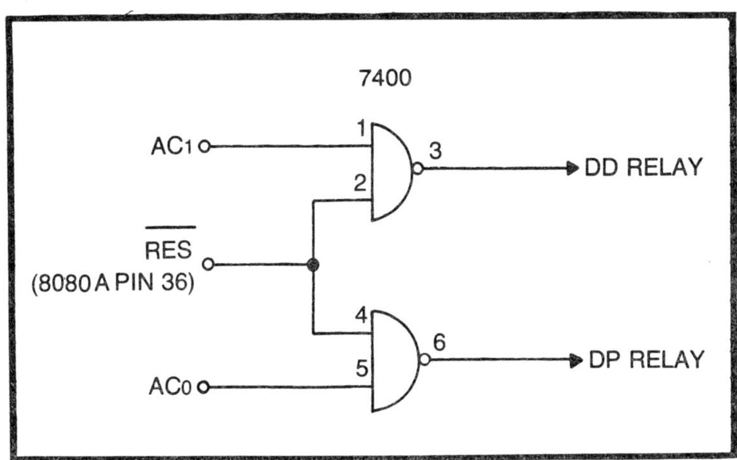

Fig. 8-4. Drive-relay control section of servodrive.

"0" and reverse motion when at a logical "1". Thus, the 6-bit output of Port AC is a locomotory word, four bits controlling steering, two bits controlling main drive. Since the unprogrammed RCU-85 produces an "all-ones" output at the present, the $\overline{\text{RES}}$ line has been gated in, so that the DP and DD relays will not energize while the RCU-85 is reset. Something similar can be done with the SD and SP relays, if they become a nuisance; be careful, though, to maintain the right polarities after the gating stage.

The first major step towards the independent operation of the robot pet has been taken. The locomotory system of the pet's mechanical frame is now fully interfaced to the digital output of the resident RCU-85 computer. When the computer is finally programmed, it will be capable of expressing its goals and decisions in visible, tangible motion. The next step is to provide the RCU-85 with some sensory subsystems, some methods of tempering its choices of action with environmental stimuli. The primary sense, Soniscan, is next on the agenda.

Chapter 9
Soniscan

The traditional homebuilt robot has a basic difficulty in navigation: it is blind. To be sure, some simple robots use photoelectric cells to track a light source, but that is far from true sight. The average robot, instead, substitutes the sense of *touch* in place of sight. To put it bluntly, it moves until it crashes against something, then reorients and tries again.

In order for a robot to act effectively in an environment, it must have adequate awareness of that environment. The "bump-and-go" sensory method is the simplest form of awareness to implement—just a few microswitches—but is somewhat crude (and sometimes detrimental to furniture). It is possible to use photocells to follow a white strip on the floor, but it would be better not to have to modify the home to that degree. Capacitive proximity sensors don't function on all materials. At the other end of the spectrum, there are semiconductor photo-sensitive arrays and even small television cameras—but the data from these are vast and require considerable brain power to interpret.

The optimum option would be some form of simplified radar or sonar which would provide only one piece of data: distance to a nearby obstacle. A beam of sound, in particular is easy to generate and slow enough to handle with inexpensive components. Such a sound-based system, termed *Soniscan* in the project, is as sophisticated in design as a bat's or dolphin's navigation, and yet is easily integrated into the simplistic sensory scheme of the robot pet.

The theory goes something like this. Sound travels at a fairly constant rate, an average of 1128 feet/sec. (343.8 m/sec). If a burst of sound is emitted toward a wall, and an echo is received *x* seconds later, then the burst made a round-trip excursion of *x* times 1128 feet, and the distance to the wall is 1128x/2 feet. Figure 9-1 illustrates this principle.

The Soniscan section of the robot pet's brain board makes use of this basic theory, with a few extra qualifications. First, the timing must all be digital for the CPU to handle it. Second, the beam must have three possible directions: forward, left, and right. And third, the beam must be *ultrasonic*—out of hearing range—for the comfort of coexisting human owners.

SONISCAN DESIGN

Let's start with the tone-burst generation. It would be simple to generate a frequency and chop it into bursts via a set of linear timer chips and such, but part of the design philosophy of the robot pet is centralization. We already have a reliable frequency source: the 8085A system clock. Assuming a crystal of 3 MHz, the 8085A runs at 1.5 MHz, and that clock frequency is available at pin 37 of the CPU. All we have to do is divide it down, with binary counters, to a good usable frequency. (If the plans in the preceding chapter have been implemented, then these counters are already in place and ready to use.)

If we feed the clock into a 7490 decade counter, a division by ten is obtained, resulting in 150 kHz. We can then direct this into a 7493 binary counter, and from pin 9 obtain a division by four, or 37.5 kHz. This is a good frequency to use: not so high as to require expensive sound transducers, but high enough to be beyond the hearing of both humans and their other more conventional pets.

A continuous beam of 37.5 kHz would interfere with echo-reception; it must be chopped into short pulses. In addition, a new pulse should not be emitted until a reasonable amount of time has been allowed for an echo to return. Finally, some sort of binary timer must be running, counting the time from emitted pulse to received echo.

A good way to meet these requirements would be to divide the frequency further with more binary counters; a slow repetition frequency would be obtained, and the counters could be read by the CPU as a numerical record of time elapsed after each new pulse.

Herein is the beauty of those 8155 I/O chips, once again. The internal 14-bit binary timer/counter can be used in a mode such that, once loaded with a binary number, it will divide an incoming frquency

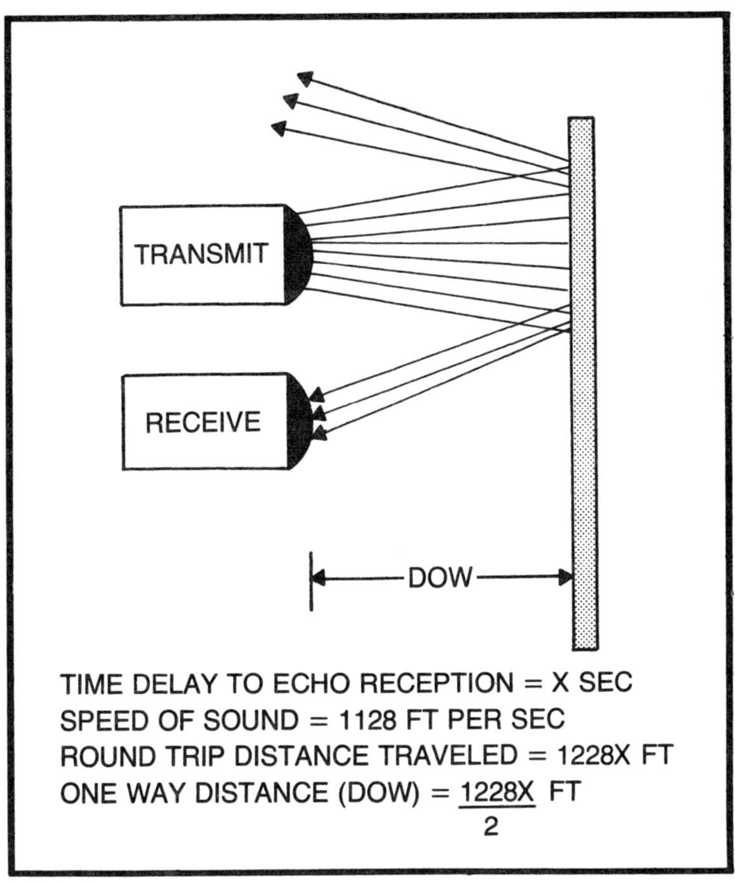

Fig. 9-1. Illustration of the concept of ultrasonic obstacle detection.

by that number. Any number up to 16,383 (14 bits of all "1,s") can be loaded into the 8155 by the CPU as a divisor.

Review the loading scheme for the 8155's timer, as illustrated in Fig. 7-6 of Chapter 7. There are two 8-bit storage bytes which the CPU can load, the lower one with I/O address XXXXX100 and the upper one with address XXXXX101. The lower byte contains the lower eight bits of the count-length number to be loaded. The upper byte contains the upper six bits of that number, plus two extra bits to tell the timer the mode in which to operate.

If the two mode bits are set to 01, the counter counts down from the 14-bit value specified, clocked by any frequency fed into pin 3. When it reaches "zero," it retriggers and counts down from the *stored* number again. The output at the $\overline{\text{TMR}}$ pin, pin 6, is a square wave, with a frequency slower than the clock frequency by a factor equal to the stored number. It is , in fact, a divided-down resultant waveform.

Now, the last counter we discussed was the 7493, which gave us a divide-by-four at its pin 9, yielding our "beam-frequency" of 37.5 kHz. There is a "divide-by-sixteen" available at pin 11 of the 7493, which gives 9.375 kHz. This can be fed into the 8155 timer via pin 3 and divided in any way desirable.

The best division would be by 256, for a few reasons. First, it yields a frequency of about 36.6 Hz, which can be used as the repetition-rate of the frequency burst. A repetition of 36.6 Hz is slow enough to allow plenty of time to receive an echo. It amounts to 27.3 milliseconds between pulses. In this much time, sound can travel, round-trip, 30.8 feet (1128 × .0273). Practically speaking, this means that the robot pet could "see" walls about 15.4 feet away, assuming its tone-burst were powerful enough.

A second good reason for a division by 256 is for the sake of the CPU. The highest number definable by an 8-bit word is 256. If 256 is loaded, then each time a pulse is emitted, the lower 8-bit storage byte of the 8155 timer counts down from 11111111 to 00000000. In order for the CPU to determine the time elapsed, it simply reads that one lower 8-bit word at the moment an echo is received. We've already said that between each pulse, sound could travel a round-trip of 30.8 feet (15.4 feet one-way to a wall). Thus, sound travels 30.8/256 feet per each increment counted down from 256, round-trip. For example, if the byte is "150" when an echo is received, then 256-minus-150 counts have elapsed, or 106. Sound has then traveled 106 × (30.8/256) or 12.75 feet round-trip. Conclusion: The wall is 12.75/2 or 6.375 feet away. (The RCU-85, of course, does not perform these calculations. It does not care how many "inches" away an obstacle is; it just wants a relative measurement to determine approximate location.)

So, then, we feed the signal from pin 11 of the 7493 into pin 3 of 8155-B. The number 01000001 and 00000000 are loaded into the 8155 registers TMR HI and TMR LO—the "01" for the proper *mode*, and the other bits amounting to the number "256" as a divisor. $\overline{\text{TMR OUT}}$ pin 6 yields our "repetition frequency" of 36.6 Hz. It is then a simple matter to feed the 36.6-Hz signal into a one-shot to gain a pulse-duration waveform; this waveform, when gated with the beam-frequency of 37.5 kHz, becomes our output burst.

How long a pulse do we want? It must be long enough to allow sufficient wave fronts to be recognized as 37.5 kHz; i.e., it cannot just let one or two waves through. Yet, it cannot be so long that an echo could return before the burst is even finished. Depending on the sensitivity of the echo-reception circuitry, a good average is a pulse wide enough to permit 16 wave fronts. 37.5-kHz waves are 26.67 microseconds apart, so the pulse should be 16 × 26.67 or 426.72 microseconds wide.

The ideal chip to use in echo-reception is the 567 tone decoder. The 567 output pin changes states when a frequency within a predetermined bandwidth is injected. For a bandwidth of 16 percent, only about 10 wave fronts are required to identify a received frequency—well within our choice of 16.

TRANSDUCERS

We have discussed transmission and reception, but not the actual sound-producing mechanism. Standard speakers will not do, requiring too much power to drive and being insensitive to frequencies as high as those with which we are dealing. However, certain piezoelectric (i.e., crystal-driven) transducers are available on the market which operate at high frequencies, use low power, and can double as transmitter and receiver.

One good source of this sort of transducer is the electronics surplus market. Certain security alarm companies make use of piezoelectric transducers in ultrasonic intruder detection systems, and they are therefore readily available on the surplus market. The pet robot prototype makes use of one such transducer, the MK-109, produced by Massa, Inc. These units are small, rugged, proven in use, and "plug-in" in design—ideal for this sort of application.

Fig. 9-2. Head of the robot pet, with ultrasonic transducers mounted in three pairs.

The beauty of piezoelectric transducers is their efficiency. Power is at a premium in our robot dog, but these transducers consume very little current—somewhere on the order of 10 to 20 mA with a 5-volt TTL square-wave input. Plus specialized transducers such as the MK-109 are made to resonate at around 40 kHz. They tend therefore to be more immune to noise and unwanted frequencies.

Figure 9-2 shows how these tranducers are mounted on the head of the robot, a new physical addition to the pet's body. The MK-109s have as a connector a standard phono jack, built in as a part of the casing. The head-piece simply holds the mating phono plugs, with the chassis available as a common ground connection. The transducers are thus removable for checking or replacement. Figure 9-3 illustrates the construction of the head-piece as it resulted in the prototype version. The central vertical angle-piece enables the head to be mounted directly on the forepart of the robot pet's sensor-frame.

The reader will note in Figs. 9-2 and 9-3 that there are six transducers in the Soniscan system of the robot pet—one pair each for front, left, and right directions. The immediate question is, "Why so many transducers?" The last thing the pet needs is greater complexity—the more parts, the more possible sources of error. This is a legitimate question, and it should be considered in a bit of detail.

In the first place, what about the triple-direction aspect? That is, why not a simple front sensor? The primary reason for the multidirectional system is for greater awareness of environment. This advantage will show up most clearly in what the pet is able to do based on surroundings. A front-sensitive pet can avoid obstacles fairly well. But a "triocular" pet can, for instance, roll parallel to a wall and then proceed to turn through a door. Also, one of the more interesting programs—the "watchdog" routine—is much more effective if the pet is able to scan a wide area, such as in the three-way version. In the end, the reader may choose to discard these niceties in favor of the simplicity of the single-direction approach. It's largely a matter of taste. For that matter, if the reader is clever, he could probably place a pair of transducers on a "rotating head" and accomplish most of the same functions of the three-way version. More power to you!

In the second place, why a *pair* of transducers? Is it not possible to use a *single* transducer, pulsing it to transmit and amplifying it to receive? This was, in fact, my intention in the original plans for the robot pet, but I ran into some difficulties. The trick is to deactivate

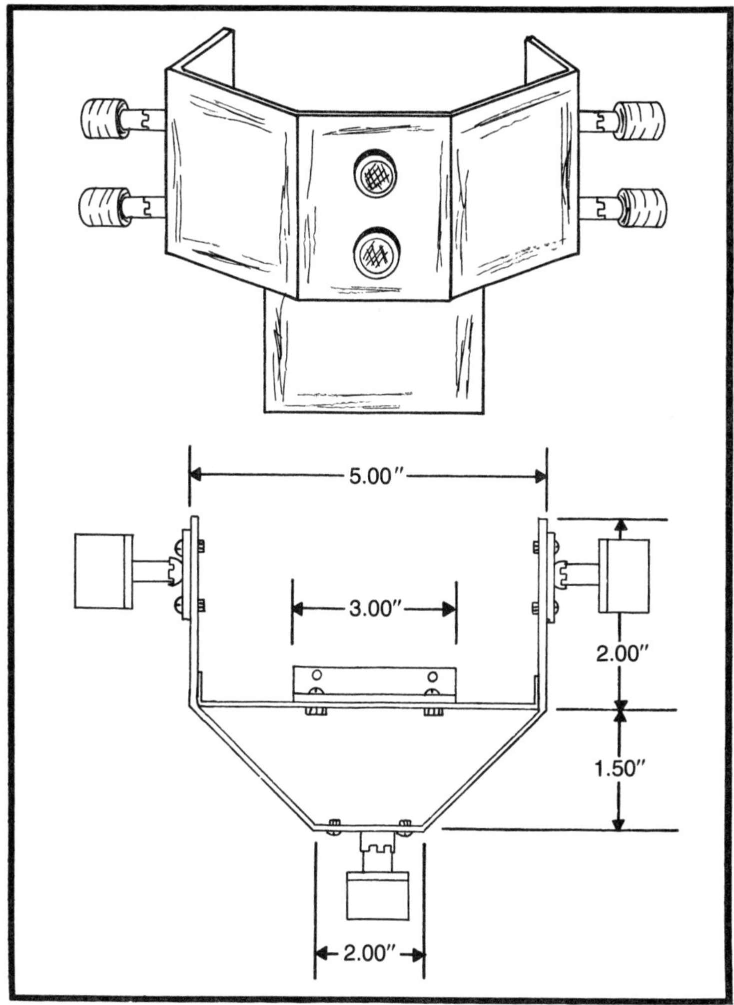

Fig. 9-3. Dimensions of the head-piece.

the reception circuitry during the transmit pulse and allow it to reactivate immediately thereafter. This takes some doing, and in the sensitive world of op amps, I encountered the stubborn demon called *oscillation*. Contributing to the problem was the way in which the transducer would tend to ring a bit after the transmit pulse, triggering the reception stage.

The single-transducer sonar system, of course, is practicable and in use in various applications today. If the reader wishes to pursue such a version in his personal pet robot project, there is an integrated circuit available which should simplify the design im-

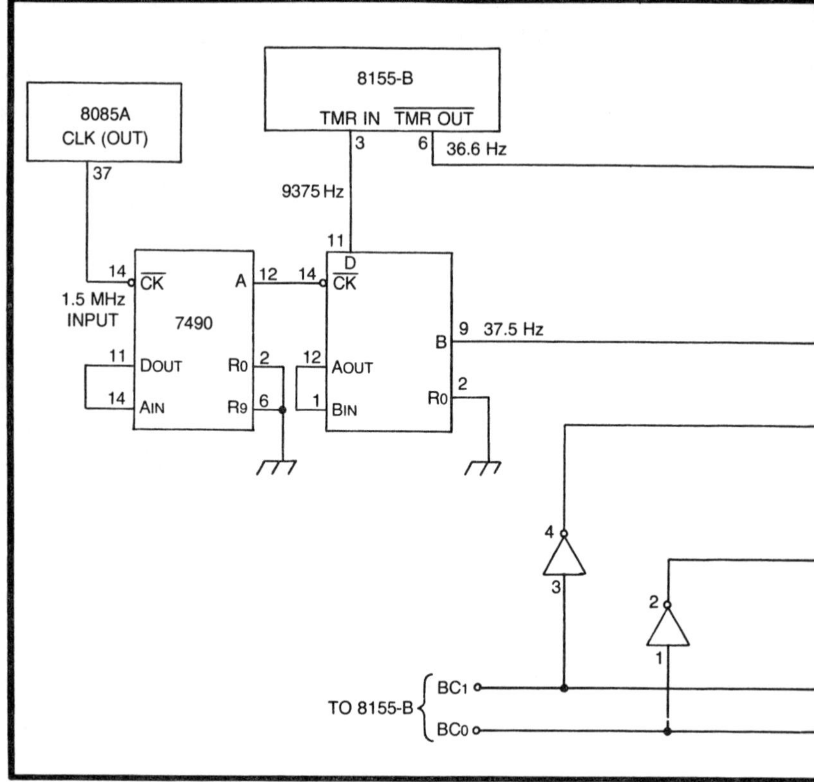

Fig. 9-4. The Soniscan transmitter circuitry.

mensely. It is the LM1812 (National), a linear circuit in an 18-pin DIP package. This single chip produces the transmit pulse, keeping its reception circuit "dead" temporarily; then it uses the same transducer to await the echo. Since the same oscillator network is used both for transmit and receive, the recognition reliability is high. The circuit can be used either for underwater sonar applications at a frequency of 200 kHz, or in air at around 40 kHz. The robot pet's MK-109 transducers would work well with this chip.

The only drawback to the LM1812 chip is the need for a few handwound inductors for the oscillator and such. For simplicity's sake in this book, I chose rather to go with the method described herein, using separate transducers for transmit and receive. The transducers do not cost much, so this should pose no financial difficulty for the Soniscan system. It may be possible, experimenters take note, to use op amp active-filter techniques to replace the hardwired inductors of the LM1812—in which case, its use would be more attractive. "Reader's choice," as usual.

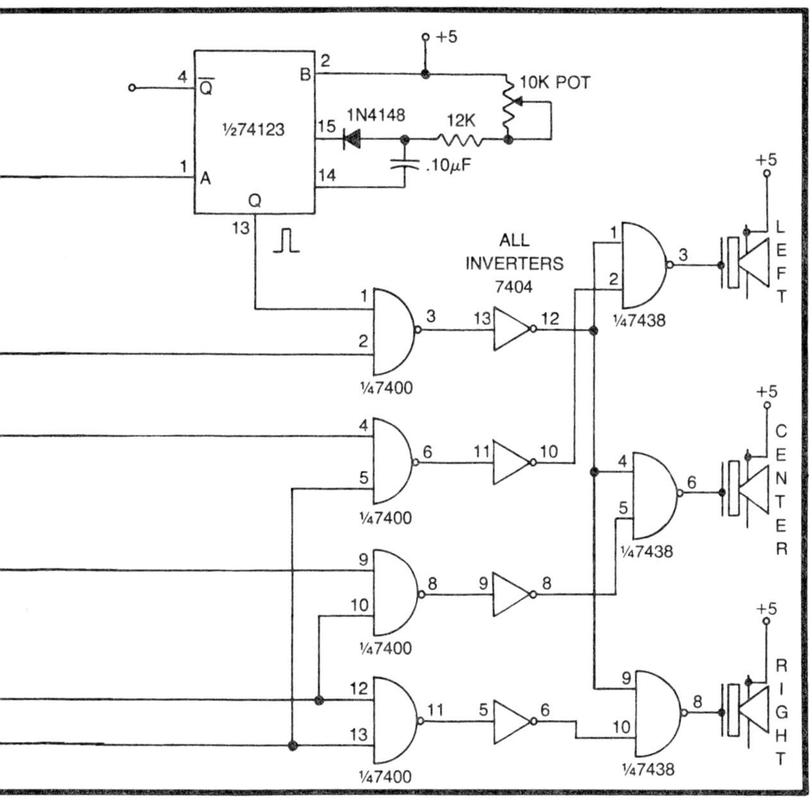

CIRCUIT DESCRIPTION

Let's take a step-by-step look at the method used in the prototype pet. Figure 9-4 gives the schematic for the "transmit" portion of the Soniscan circuit. As we have already described, the 7490 and 7493 produce resultant squarewaves of 37.5 kHz and 9.37 kHz. The latter wave clocks the timer in 8155-B, which in turn produces a divided-down waveform of 36.6 Hz. This wave triggers a 74123 one-shot on the positive edge, producing a pulse with a width of about 430 microseconds. When this pulse is gated with the 37.5-kHz signal derived at pin 9 of the 7493, the result is a recurring "burst" of frequency, having about 16 pulses. The number of pulses per burst, of course, is a function of the timing of the 74123; thus, the trimmer potentiometer shown with the one-shot can be used to find the optimum number of pulses for reliable use.

This series of tone bursts is fed into the driver-circuits of all three transducers; a simple two-to-four-line decoder is used to enable only one of the three. This decoder is controlled by bits BC_0 and BC_1

of 8155-B. "00" enables none of them; "01" enables the left sensor, "10" the front sensor, and "11" the right sensor.

The driver circuit has to be unique, because of the unique characteristics of the piezoelectric transducers. It is not enough simply to pulse the transducer and then leave it to float at the high-impedance state;the crystal responds with an almost capacitive tendency to remain charged. And so a simple driver transistor is not really sufficient—the transducer ought really to be "forced" from ground up to +5 volts, then down again, using a complementary pair of transistors. Sufficient for the purpose is the 7437 NAND buffer IC as shown; the output is capable of about 50 mA, which is plenty for the efficient transducers.

The *receive* portion of the Soniscan system, as depicted in Fig. 9-5, is more delicate. Each of the three transducers has its own "receive" circuit, although the figure only shows a typical circuit with the final stage which unites all three. Essentially, each transducer has a two-stage amplifier, making use of the two op amps in a 558 IC. The output of the amplifier-stage is then sent to a 567 tone decoder, configured so as to identify a 37.5-kHz tone. When the echo pulse is received, the output of the 567 drops to zero volts; this changes the state of the following NAND-gate. The NAND-gate output thus informs the RCU-85 of the reception of an echo—provided that it is not first "headed off" by the intervening gate. This latter gate does not permit the RCU-85 to mistake the *transmit* pulse for an echo, by gating in the inverse of the 74123 waveform from the transmit section.

The alignment of this receive section is touchy, indeed, but it is not as bad as it might first appear. In our favor is the fact that the circuit is somewhat noise-resistant. The reason is two-fold: the transducers themselves have a low sensitivity to frequencies other than about 40 kHz; and the 567 decoders require at least 10 pulses to identify a tone. The alignment procedure will be dealt with shortly.

The transmit portion is simple enough to wire up on the brain board. The 7490 and 7493, if not already wired during the last chapter, are first. Then the 74123 is installed. Use another 16-pin DIP socket for the resistors and capacitors which specify the pulse-width, except for the trimmer potentiometer, which can be mounted nearby. Finally come the gates and inverters for the two-to-four-line decoder and such. The 7437 NAND buffer IC should be mounted on a small card up in the "head" of the robot, near the transducers. Then the lines carrying the transmit pulses and select-levels can be run from the previous chips on the brain board, through the proper pins of connector "I," and up to the head. The reason for

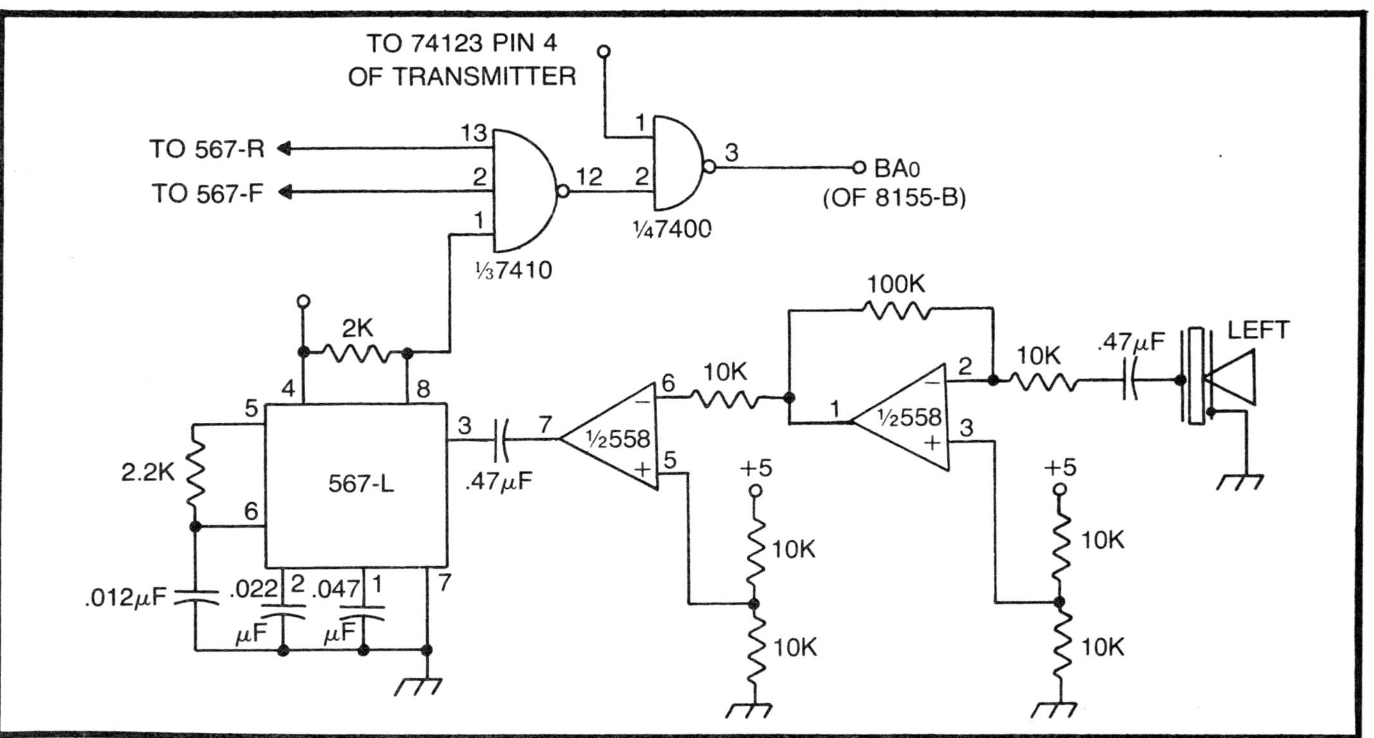

Fig. 9-5. Representative Soniscan receiver circuitry; the uppermost two gates are common to all three receivers.

this approach is to keep from having significant driving current (and pulsed, at that!) flowing among the sensitive ICs of the brain board. In fact, the four low-level lines should be routed as far away as possible from other lines anyway, just in case.

The receive circuitry, to take it to extremes, is almost entirely located in the head, not the brain board. A small card (separate *especially* from the 7437 buffer chip) is mounted in the head, bearing three identical circuits. Each circuit will require three DIP sockets: one for a 558 dual op amp and a few parts, one strictly for parts, and one for the 567 decoder with parts. Trimmer potentiometers can be mounted and used, or discrete values can be substituted experimentally until alignment is successful. The three levels from the 567s are routed back to the brain board via connector "I," and are gated there as per the figure.

In wiring both the transmit and receive portions in the head, use short, efficient wire-wrap leads to help stabilize the system. In addition, it's a good idea to use some light coaxial cable to connect the six transducers. Note the word "light"; the currents are fairly low, so don't choose a heavy-gauge coax that will be difficult to work with.

ALIGNMENT PROCEDURE

Once the whole Soniscan system is installed, it is time for alignment. Since the RCU-85 is far from programmed, it will be necessary to sort of "sidestep" its part in the system. This is done by providing some external source of a slow frequency to trigger the 74123 in the transmit portion, and also externally controlling the select-lines which enable the individual transducers. If you have a square-wave generator, the first step is easy: simply remove the 8155-B chip to avoid interference and then feed a 36.6-Hz TTL-level square-wave into the 8155-B socket at the pin 6 location. This in turn will trigger the 74123 normally. If you don't have a generator, the circuit in Fig. 9-6 will provide the proper waveform; its output, as above, can be fed into the empty 8155-B socket at the pin 6 location. The select-lines can be controlled simply by grounding the empty 8155-B socket at the locations for pins 37 and 38, simulating a 01, 10, or 11 for the three transducer angles.

Make the substitutions as noted above, and power-up the circuit. With an oscilloscope, monitor the relevant waveforms. Check for the 37.5-kHz wave and the 430 microsecond one-shot pulse. Examine the result of the gating of these two. Then look for this signal out at the individual transducers, each one as it is enabled by the proper grounding at the empty socket of the decoder circuit.

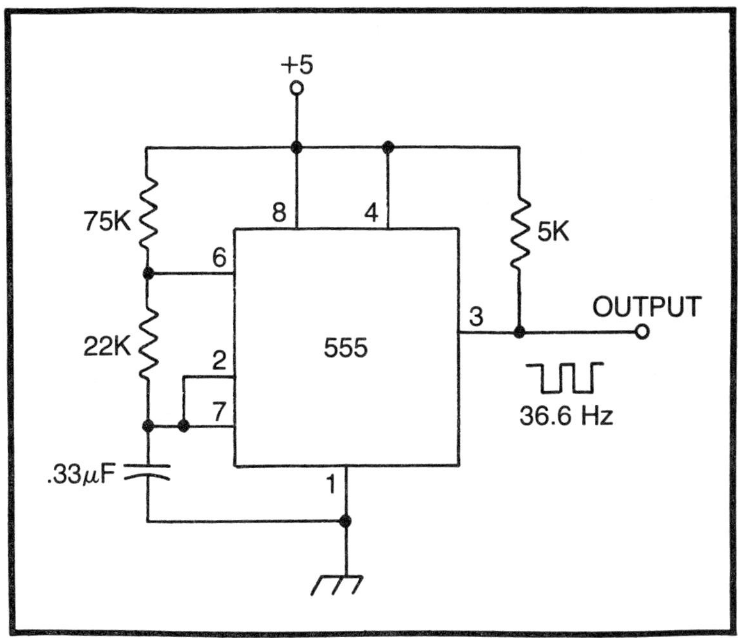

Fig. 9-6. TTL square-wave generator for alignment of the Soniscan system.

Next hook the oscilloscope to the input of the first op amp of the receiver circuit of the pair now being used in transmitting. Place a flat object parallel to the pair of transducers, or face them toward a wall, only a few inches away. Turn up the sensitivity of the oscilloscope until the received waveform is visible—it will be in the millivolts at best. (If there is a strong 60-Hz wave superimposed, do some grounding of equipment to eliminate it.)Then move the oscilloscope to the output of the first op amp, pin 1 of the 558. Adjust the alignment potentiometer (or substitute values) until a strong, amplified version of the waveform is visible. Then move on to the output of the second op amp, at pin 7, and repeat this process. If possible, use a plastic aligning rod to prevent body capacitance from disturbing the circuitry.

Now, slowly remove the flat object or move away from the wall. As this is gradually done, the burst seen on the oscilloscope should separate into *two* bursts; that second, much weaker one is the echo from the wall or object. Move the oscilloscope to the output of the 567 tone decoder. Align the 567 so that a clear negative-going pulse is visible with as little "chatter" as possible. As the object or wall advances or recedes, this should be visible from the 567 output: two separate pulses should be seen to diverge and converge. Again, the first is the direct transmit-pulse picked up due to the proximity of the two transducers; the second is the echo.

Repeat the aforegoing alignment procedures with all three sets of circuitry. Then observe with the oscilloscope to see that the 567 outputs are indeed bringing about the change of state in the NAND gate on the brain board. The output of the second NAND gate, the one in which the 74123 pulse is involved, should be *one* pulse, not two; the first, direct pulse should be altogether removed by the gating, leaving only the echo. If this is not the case, then ringing in the transducers or instability in the receive circuitry is causing a lengthening of that first pulse *past* the predetermined pulse-duration. If this is the case, and the ringing cannot be stopped by further alignment, solve the problem by using the second one-shot in the 74123 package. Wire it like the first one, triggering it from the same input; hook its output to the NAND gate in question, in lieu of the first one. Adjust it, using a potentiometer or by selecting values, so that its output is a bit *longer* than its first counterpart. Watch the output of the NAND gate, and adjust the one-shot until every last trace of the first received pulse is gone.

SONISCAN IN OPERATION

The software support for the Soniscan system will be dealt with in a later chapter, but here is an idea of what is involved. In a certain program sequence, the RCU-85 decodes that it wishes to survey its surroundings. The transmit circuitry has been producing pulses all along, but these pulses have not been allowed to drive any of the transducers. And so the RCU-85 selects the left transducer by outputting a "01" to port BC. This permits the transducer to be driven, and a "burst" is emitted. A short time later (all the time during which the RCU-85 has been patiently waiting), the receive circuitry picks up the echo, and notifies the RCU-85 of the fact via pin BA_0. Immediately, the RCU-85 interrogates the internal timer/counter of 8155-B, which all of this time has been counting away the moments since the transmit-pulse. Once the number has been stored away, the RCU-85 selects the front transducer, and then the right transducer, each time repeating the procedure. Now it has stored in memory three numbers, each of which gives a relative idea of the proximity of obstacles. The higher the number, the closer the obstacle. If the number is virtually zero, the obstacle is safely out of the range of the Soniscan system. The entire process of Soniscan updating occurs in such a short time that it can be done often, in the midst of more important tasks, without significantly delaying these other tasks.

What drawbacks are there to the Soniscan system as described? One is the limited range. Because the driver circuitry is limited to TTL voltages (i.e., +5 volts), the output power of the transmitter is severely limited. It has been sufficient in most cases, though, and a range of only three or four feet is still better than the "zero-range" of bumper-contact systems. The serious experimenter may wish to play with more powerful transmitter circuits, but they will probably require using an auxiliary battery furnishing a higher voltage for the transducer drivers. This may be practical, since almost any such battery would last long under such minimal loading conditions (only some 20 milliamps for an occasional duration of 430 microseconds). A 9-volt battery might serve. Remember, though, that the higher voltage will require the design of a discrete driver circuit, since the TTL NAND buffer used herein will not tolerate these voltages.

Another problem is the sensitivity of the receive portion of the circuit. If care is not addressed to the placement of parts and the running of leads, you will be forever plagued with outbreaks of oscillation. There is definitely room for development in the sophistication of the circuit. The experimenter may wish to try higher grade op amps which can be externally compensated for frequency. Or he might play with dampening capacitors in crucial locations. Then there's always the option of a totally discrete transistor amplifier. The final circuit is not as important as the concept behind it: reliable, moderate-distance proximity detection.

The Soniscan circuit is among the features unique to the pet robot system as far as average robots go. Once it is completed and functional, the projects steps out from among the run-of-the-mill and takes on a sophistication rarely seen in the home-experimenter's workshop. In the next chapter, another unique feature will be integrated into the pet, one which will advance it one step closer to the goal of "life simulation."

Chapter 10
Excom

The robot pet, like most pets and *unlike* most robots, is a highly independent creature. As we will see later, the standard software package for the pet robot enables it to roam around freely, selecting from a number of available subroutines, with a minimum of supervision. In this sense, a robot *remote control* circuit is quite unnecessary, for the pet is wholly capable of watching out for itself.

And yet, some sort of nominal external communication and control is desirable. A pet that totally ignores its owner is not a pet. The owner has certain ideas he needs to convey to his pet—such as "Stop barking! It's almost midnight!" or "Don't sit in front of the TV set!" In other words, the average pet has a list of usually-acceptable practices and habits—but on occasion, a particular practice is *not* acceptable. Then the master-to-pet relationship must come into play and override the pet's free choice.

So it is with the robot pet. Ordinarily its practices are quite harmless, since its entire subroutines repertoire is user-delineated. But some sort of override capability is necessary in special situations. The robot pet may need a "Be quiet" command, or a "Shut down for programming" order. For that matter, a set of diagnostic commands would be useful, commands which put the pet robot through preprogrammed sequences to demonstrate its abilities.

There are a few approaches to this necessary addition to the project. The easiest is to build a direct remote control device which, via switch settings, directly overrides the brain board and "seizes control" of the pet's body. A variation with a bit more finesse is a control box which jumps the brain's programming to special subroutines. This is certainly a practical sort of concept, and the reader may choose to pursue it. But there is another direction, dictated solely by the desire to approach "life simulation" to an even greater extent. This other direction is the incorporation of *speech recognition*.

The advantage of some sort of speech recognition is chiefly aesthetic. It is much easier to think of the robot dog as a life-like pet if it does not require a little gadget for communication. And when you most want to command the robot pet, it usually seems that the all-important control box is sitting in some other room somewhere. Your voice, however, goes where you go. But the major difficulties of speech recognition lie in its complexity. Speech recognition boards are complex, and presently they carry a price tag reflecting the fact. And small wonder: The factors involved in speech and language, particularly that formidable personage know as Grammar, are many and difficult to isolate and classify.

It is clear, then, that if vocal communication with the pet robot is to be practical, it must be in some other language than standard English. The resultant person-to-pet language must be a compromise, yielding concessions to the modes-of-thought of both man and robot. Each must, in short, learn some "translation."

The result of this conflict in communication is a hybrid language which we'll call *Fredian*. Fredian is derived from FREquency DIfferential ANalysis, which is the process by which Fredian is translated into meaningful data. Fredian is a spoken language, as a concession to human practice and yet it is binary in format, in deference to robotic custom. The owner can learn to speak and understand Fredian with a little practice. The robot pet, though, requires a special circuit and program routine to perform its translation. The circuit and routine will be referred to as *Excom* for EXternal COMmand.

FREDIAN GRAMMAR

The principles behind the Fredian grammar are straightforward. The pet robot thinks in terms of binary chunks of data known as *bytes*, 8-bit binary words. As it turns out, for most applications, the most binary data needed in Excom situations is a 4-bit chunk (a "nybble," if you will).

On the human end of things, the most easily distinguishable factor in speech is *pitch*. Though not many people have the musical ability to generate and recognize perfect pitches, most can *compare* two widely differing pitches and decide which is higher in frequency. Pitch is more simply handled by electronic circuits, also, as opposed to the more elusive elements of speech such as consonantal sounds or timbre. It is a relatively simple task to reduce frequencies to numbers, and then to compare those numbers. It becomes clear that pitch-communication is a method which both robot and man can translate.

A robot, if built with accurate-enough reference components, can listen to a frequency and immediately identify it in terms of cycles per second. Humans are not so constructed, but though they are incapable of exact *identification* of audio frequencies, they are quite apt at relative analog *comparison*. And so, a part of the "grammar" of Fredian must be the supplying of a *reference frequency* to act as a standard for comparison. In other words, each word of Fredian spoken by man or robot must contain a primary reference pitch, to be followed then by *data pitches*.

Another inherent superiority of the robot's frequency recognition is its *resolution*. Given a reference frequency of 300 Hz, a simple circuit could detect a comparative difference of, say, 10 percent, and thus be able to recognize data pitches varying from 270 Hz to 330 Hz. Specific human owners may or may not be able to distinguish pitches so close, and if they can, it is certain to require some concentration and ideal ambient noise conditions. And so Fredian must make use of frequencies of a fairly wide comparative difference, to allow for easy translation by the human speaker/listener.

The robot has a few "weaknesses" of its own, to which the human must make concessions. The first is in terms of data. Human beings operate in the realm of *analog*, and thus their languages, though imprecise, contain much data. Human languages attempt to make up for imprecision by the sheer multitude of information which an analog set of sounds can convey. Computers, on the other hand, operate in the *digital* realm. The transfer of digital data is inefficient with respect to time and space, but compensates by the pure *exactitude* of the data. Digital data words mean one-and-only-one thing at a given time, and therein lies their strength.

The person speaking Fredian must orient his speech to the basically binary format of the robt. Fredian words mean one-and-only-one thing, because they are binary in construction. The owner of the robot pet, in short, must program the pet to understand certain Fredian code-words, and then memorize the codes himself. The trick is in learning them as what they actually are: 4-bit binary numbers.

A final bit of grammar to relate Fredian to man and robot has to do with the spacing of data. The robot requires something to help

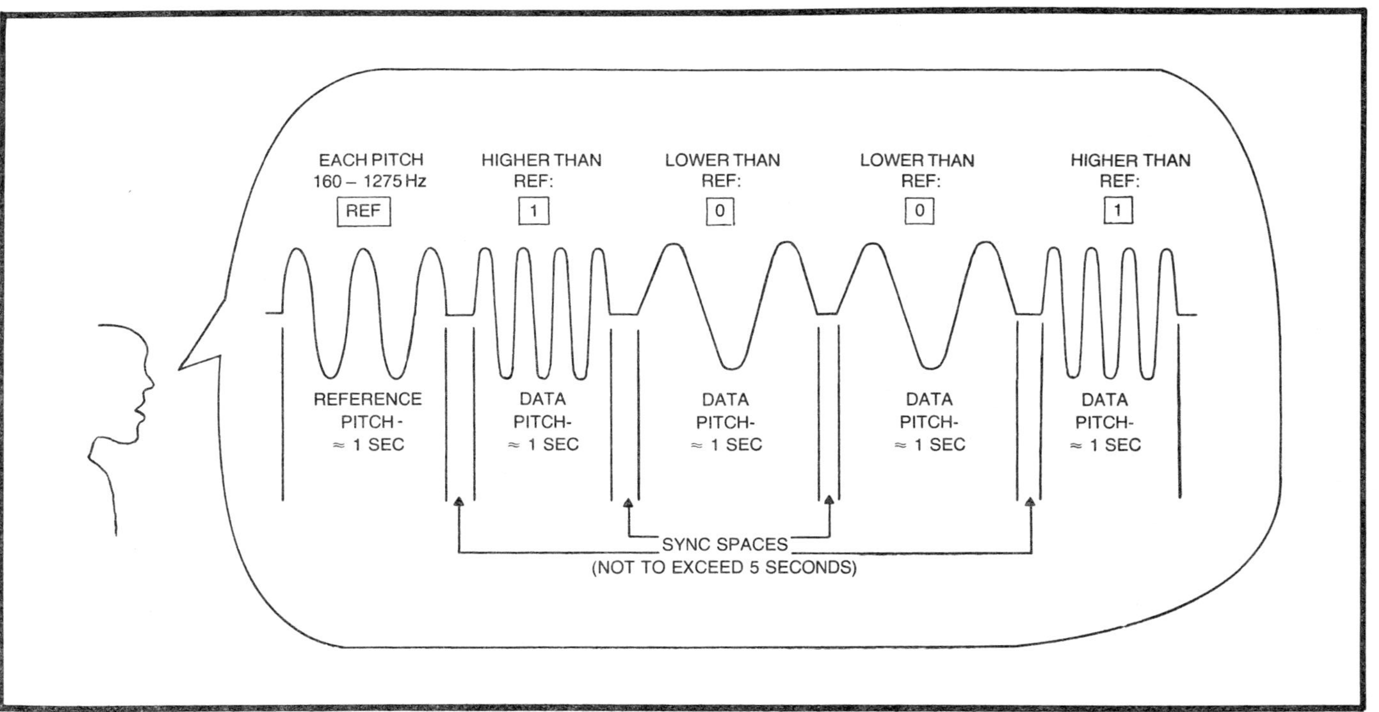

Fig. 10-1. A summary of the Fredian grammar.

synchronize the data bits coded in Fredian form. So the human speaker must place silent spaces between each *data pitch* he speaks. On the other hand, the human being cannot be expected to *time* his spaces with any real precision. And so the robot must place only a maximum tolerance on the length of the space, allowing considerable freedom to time delay on the part of the human. No exact timing restrictions should be placed on the length of the pitches, either.

Figure 10-1 provides a sort of summary of the Fredian grammar. The speaker emits a pitch anywhere between 160 and 1275 Hz, a range which easily contains the frequencies people speak and hear. The pitch need not be of a tightly-controlled length, though a one-half to one second duration is a good guideline. The listener hears this pitch and remembers it as a *reference pitch* with which to compare succeeding pitches. The speaker allows a silent space, called a *sync space*, not to exceed five seconds in duration. Then follow four data pitches (with intervening sync spaces), determined according to the following definition:

For binary data "1", data pitch is higher than reference pitch.

For binary data "0", data pitch is lower than reference pitch.

The data pitches may be as much higher or lower than the reference pitch, as long as it is within the 160-to-1275-Hz range already mentioned. No more than four data pitches are permitted. In ordinary usage, only one word of Fredian is sufficient to convey a given command. There are some possible two-word commands, though, in which case the whole process is repeated a second time. This possibility is discussed later.

The Excom circuit and related programming is designed to interrupt the pet robot in its normal free-running course. During the interruption, the Excom subsystem stores the reference frequency heard from the owner, compares each of four subsequent frequencies to it, and thereby derives a 4-bit "nybble" of data to use in selecting a preprogrammed course of response. Because of the limitations of a 4-bit number, only 16 possible words exist in the Fredian language, and thus only 16 possible responses are open to command via Excom. (An 8-bit version could have been used, but it would have been unwieldy for the human speaker to handle.)

Figure 10-2 gives the physical portion of Excom, the receiving circuit. A standard, inexpensive crystal lapel microphone is used as an input. These have a limited frequency response range, but that is just as well, since this will aid in prohibiting the consideration of extraneous noises. The microphone is hooked to a transistor buffer stage for preamplification. The output of this buffer stage goes to the input of an LM311 voltage comparator. The 311 comparator effec-

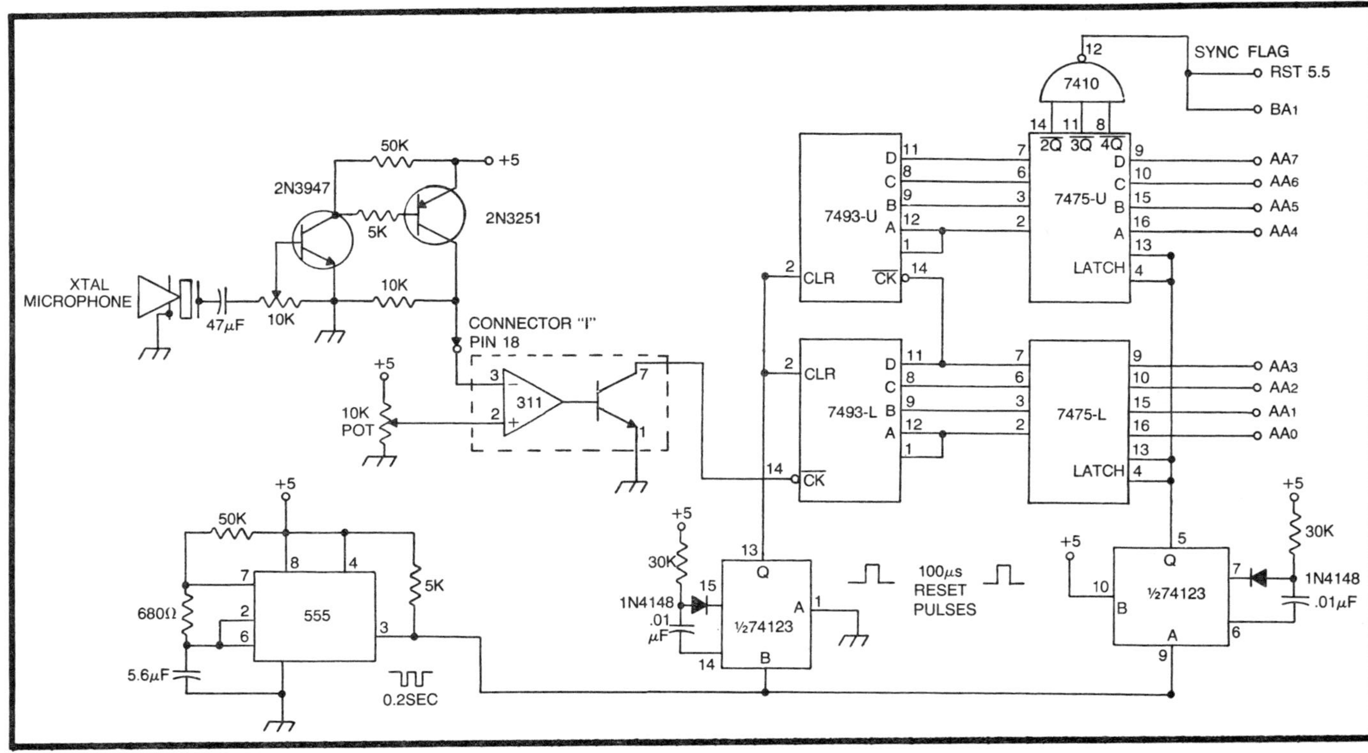

Fig. 10-2. Excom receiver circuitry.

tively converts the input frequency from a sinusoidal waveform to a TTL square wave. This enables us to deal with input pitches in terms of *zero crossings* using standard TTL counter circuitry.

That square wave is then used to clock a pair of 7493 4-bit binary counters. The 7493s will output an upward-counting 8-bit number at their A, B, C, and D pins. This output is fed into a pair of 7475 4-bit latches. Every two-tenths of a second, the latches are clocked and the counters are reset to zero by a special timing circuit. The timing circuit consists of a 555 timer, to provide the proper intervals, and a 74123 dual monostable multivibrator, to produce the pulses for clocking and resetting. The end result is a dual-7475 latched 8-bit output which gives a continuous record of input frequency as sampled every two-tenths of a second. (The actual frequency in hertz would be five times the output of the 7475, but that is irrelevant. All the robot pet needs is a relative measurement.)

The final circuit component is the *sync flag*. A 3-input NAND gate (part of a 7410 triple NAND package) is hooked to the uppermost inverted-output pins of the 7475 latch as shown. As long as the dual-7475 output is *no greater than* binary 00011111 (decimal 31), the inverted inputs to the NAND gate will each be "1", and output of the NAND gate will be "0". But if the dual-7475 outputs any number from binary 00100000 to 11111111 (decimal 32 to 255), the NAND gate output will swing to a "1". To make it clearer, as long as the input frequency remains *less than* 160 Hz, the dual-7475 output will be less than 00100000 and the gate will be at "0". But *any frequency* from 160 to 1275 Hz will trigger the gate to a "1".

This sync flag is fed to two places. It is hooked first to the RST 5.5 input of the 8085A CPU chip in the RCU-85 computer portion on the brain board. The RST 5.5 pin is internally wired such that, when it is raised to a "1", the normal 8085A program flow is interrupted and the program jumps to a specific location in program memory. At this point in memory would be the proper Excom routine to permit decoding to the Fredian words. The sync flag is also fed to port BA_1 of 8155-B in the RCU-85 unit. In this way, the status of the flag can be watched *after* the interrupt routine has already started and the RST 5.5 pin has been disabled by the program.

Figure 10-3 shows the basic program flow to the Excom subroutine when a frequency has been detected. (It is shown in block form since we have not yet discussed the necessary programming language for the RCU-85.) The 8-bit dual-7475 output is read from the circuit at port AA of 8155-A. The very first frequency number is then stored in a memory location as a reference pitch. The program waits for the sync flag to drop to zero, indicating a silent sync space.

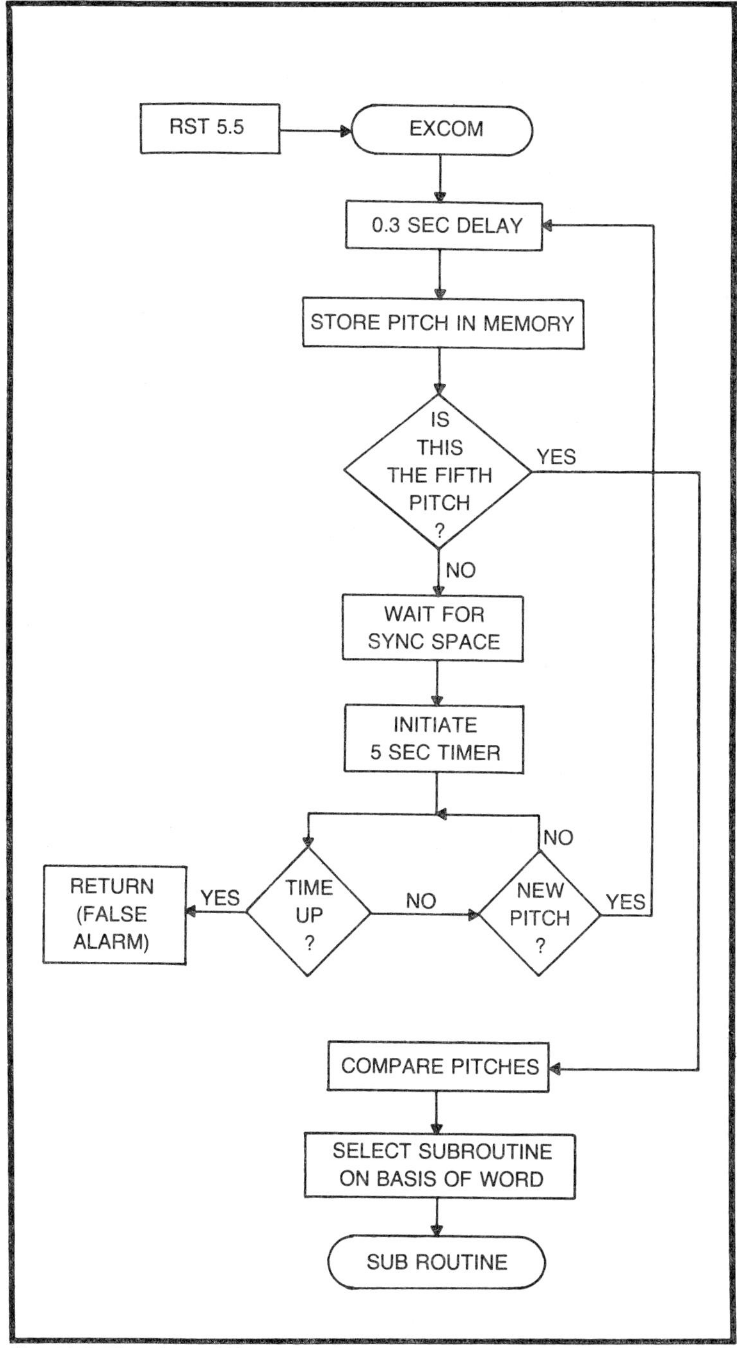

Fig. 10-3. Flowchart of Excom routine.

If no subsequent pitch is received (i.e., if the sync flag doesn't go to a "1" again) within five seconds, the first pitch is assumed to have been a spurious burst of noise, and the program flow returns to the point at which it had been interrupted. If the sync flag *does* go high again, the next four subsequent frequency-numbers are read and stored in memory locations. Then the decoding takes place: the four pitches are compared to the reference pitch number, and the results are used to form a reconstructed 4-bit nybble of data. This data may then be used to select one of 16 subroutines as a response to the command.

The power of the Excom system is formidable. The subroutine accessed by the Fredian word may be anything from a mere *cease motion* response to a fullblown multi-step sequence of activity.

One possibility with which to experiment is a special *two-word* Fredian command as pictured in Fig. 10-4. The first Fredian word accesses a subroutine labeled *Next*. The Next subroutine is designed, simply, to read in and decode a *second* Fredian word and use those 4 bits to determine some course of action. For instance the new word might be a code number for one of 16 persons who own the robot pet. The pet could be programmed to respond with "happy barking," "suspicious growling," or "fearful fleeing" based on the identity of the person. In other words, the *first* Fredian word would mean "My name is...," and the second word would be the *name code*.

CIRCUIT CONSTRUCTION AND TESTING

The Excom circuit can be wired virtually anywhere on the brain board. It will require about nine 16-pin DIP sockets: one each for the 311, 7493s, 7475s, 7410, 555 and 74123, and one for miscellaneous components. The microphone is mounted on the sensorframe behind the head, facing upward so as best to receive vocal signals from any given direction. The transistor buffer stage is mounted right at the microphone on a small piece of perforated board. The output is then routed to the 311 comparator on the brain board through pin 18 of connector "I."

As is the case with most of the higher-level circuits of the pet robot, the Excom circuit is largely "untestable" until the RCU-85 computer is up and programmed. But the basic timing pulses and levels can be checked once the circuit is wired onto the brain board.

The first place to check is the input to the 311 comparator, at pin 2. A vocalized pitch from a fair distance of few feet should result in a sinusoidal waveform at this point. If the buffer stage after the microphone tends to break into oscillation, some substitution of resistor values may be necessary. Next check the 311 output, at 311

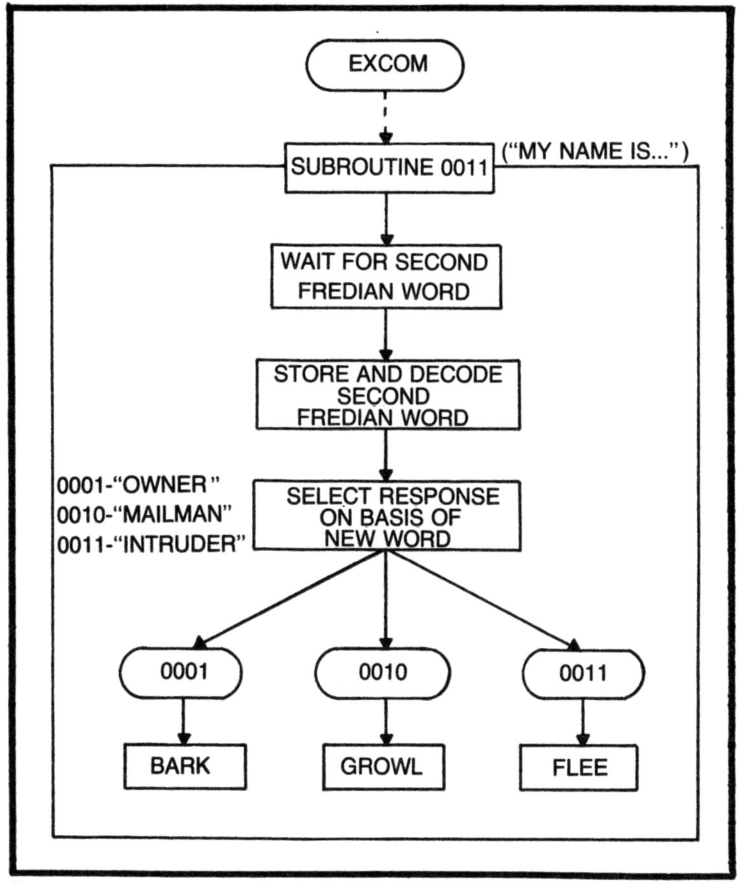

Fig. 10-4. Possible "two-word" Fredian command flowchart.

pin 7 or at 7493-L pin 14. Your oscilloscope should detect a clean TTL 0-to-5-volt square wave upon inputting a frequency to the crystal microphone.

Next, check the 555 output, at pin 3. There should be a negative-going pulse generated every two-tenths of a second—that is, at a frequency of 5 Hz. Then examine the 74123 outputs. Pins 13 and 5 should yield a 5-Hz positive-going pulse train, but the two trains will be slightly out of phase. Figure 10-5 shows the relationship of the timing pulses. They are timed to latch the 7475s first, then clear the 7493 counters next, to prevent any loss of data.

The sync flag is easy to test. With *no* input frequency, the NAND gate output should be "0". Any frequency from 160 Hz to 1250 Hz, though, should drive the "flag" to a "1". Frequencies above 1250 Hz may result in "0" or a "1", depending on the case, but the

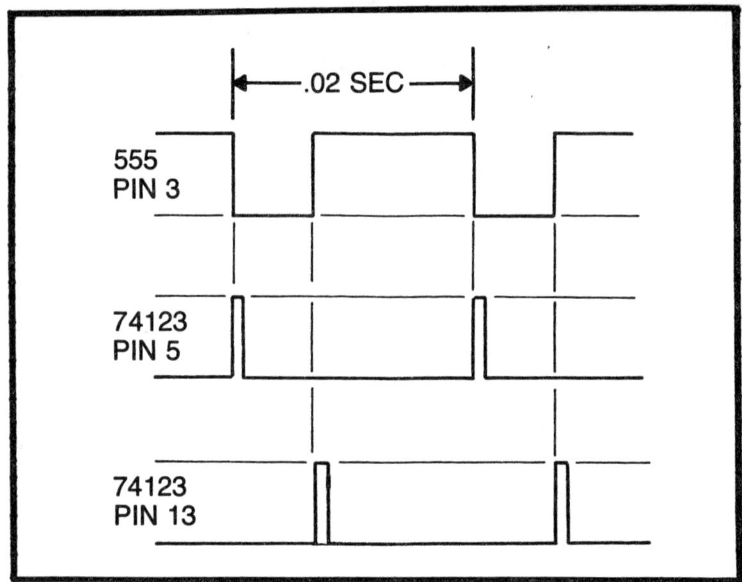

Fig. 10-5. Major waveforms for Excom receiver circuitry.

average voice is well below those pitches. Plus the crystal microphone will tend to limit sensitivity to frequencies in the human voice range.

A good final test would be to read the 8-bit output of the 7475 latches. But this is difficult to do on a pin-by-pin basis while vocalizing a frequency. Now, then, might be a good time to assemble a handy debugging tool called a *status display device*. Such a troubleshooting aid, illustrated in Fig. 10-6, is simply an 8-bit LED display with two 7416 hex inverters are flexible leads which can be temporarily attached to any point in a circuit which requires logic monitoring. Any of the RCU-85's 8-bit ports can be examined using this simple device. Naturally, it isn't helpful in the case of rapidly-changing waveforms, but it enables the experimenter to keep an eye on eight occasionally-changing levels at a time.

In testing the output of the 7475 latches, the 8155-A can be removed from its socket, and the eight leads of the display device can be inserted into the socket pin locations corresponding to port AA. The eight LEDs should read out the latched frequency based on a vocalized input pitch. In ordinary, low-level ambient noise, the display should read out a binary number less than 00011111. Vocal pitches will result in a displayed number between 00100000 and 11111111, at which time the sync flag should go to a logical "1". Expect the readout to flicker some as any vocalized note will vary in pitch somewhat. But if the display indicates total instability, there is a serious problem, probably either with the input buffer stage or the 74123 waveforms.

Try producing Fredian words experimentally into the Excom circuit, along the guidelines set in Fig. 10-1. Vocalize a pitch for the *reference frequency*, using a vowel sound such as "ah" rather than a "hum." Choose a pitch which is in the middle of your vocal range, so that it is fairly easy to produce pitches above and below that first reference frequency. Give the note a duration of about one second and then cut it off *abruptly*, so that there is a clean break to trigger the sync flag. Pause about a second, than vocalize a second pitch, higher than the first one, to represent a binary "1". Treat it with the same timing guidelines as the first pitch. Then "la" a third pitch, lower than the first one, to represent a binary "0". Follow up with a fourth and fifth pitch to complete the Fredian "word." If you choose a good first pitch, and produce the data pitches significantly higher or

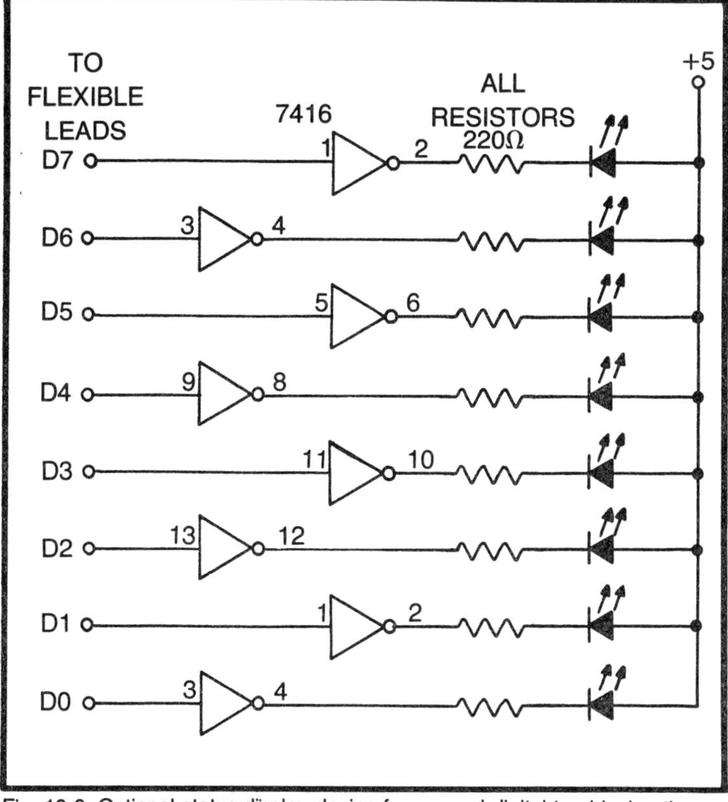

Fig. 10-6. Optional status display device for general digital troubleshooting.

lower than it, you should have no difficulty. Monitor the status display device and the sync flag to see how the Excom circuit tracks the input.

It will not be until Chapter 16 that Excom will come into full working capability, or it is then that the robot pet is programmed to store the pitches, compare them, and apply them. Until then, consider what this simple TTL circuit accomplishes. It is a vocal interface, analyzing a vocal input and making it interpretable to a microcomputer. True, it is not a fullblown speech recognition unit, identifying English vocabulary. But neither is it simply a voice-triggered input, responding merely to the presence of noise. It is a mediating approach, simple and inexpensive to implement, yet complex in terms of possible applications.

Now, wouldn't it be nice if the robot pet could respond vocally as well? The circuit that will give the pet its "bark"—the Audigen circuit—will be explained in the next chapter.

Chapter 11
Audigen

When young children learn about animals, more often than not, they associate some characteristic sounds to them. They may not know much about a cow, but they know that it goes "Moo." We carry these associations with us into adulthood whether we realize it or not. To us, an animal is *identifiable* by its sounds, and we are not comfortable with a pet that is silent. Indeed, we rely on the sounds of an animal to determine its state of health or happiness.

The pet robot has reached the point in design where it can be spoken to by the owner. Now it is time to give the pet a distinctive vocal system of its own, a means for it to respond audibly to stimuli. The circuit will be termed Audigen, a shortened form of AUDIo-GENerator. Once this subsystem is added, the robot pet will be able to alert its owner to important situations, such as low battery voltage, or simply to make everyone aware that it is "coming through." It is even possible for the pet robot to communicate in specific terms, making use of the Fredian language described in Chapter 10. To that degree, the Audigen is not simply an add-on "for show"; it is a necessary step toward our design goal of life simulation.

DESIGN CONSIDERATIONS

What options are open in providing a robot with a voice? The range of choices is broad, somewhat similar to the question of voice recognition. On the simplistic end, a robot can have a one-sound device: a bell, a buzzer, an electronic tone or "beep." On the complex end, there is speech synthesis (or for dogs, "bark synthesis," if you will), using tape recordings or some form of digitized speech. Speech synthesis is a complicated approach, usually requiring a good deal of memory storage; single "beeps" are not very useful. So let's see if we can strike some sort of balance.

A pet usually has a few different pitches which it can utter, representing moods of anger, calm, or alarm. The robot pet ought to be able to express this sort of range of "emotion," at least so that the owner can tell if it is normal or not. Pitch variations would be necessary if the pet robot were to communicate in Fredian, as well. Animals also convey their meaning by repetition, from a few cautious barks to rapid yelping. The pet robot shoudl be able to "bark" once or in various speeds of repetition, to indicate its intentions. Lastly, variety of expression is capable through the length of a sound, be it gradual or clipped.

From the point of view of the RCU-85, the resulting circuit should be easy to control. The final Audigen subsystem should have a certain amount of "intelligence" of its own, with the RCU-85 merely deciding on the advisability of a "bark" at the time. In other words, the processor should not have to handle all the timing and pitch production in software; this should be delegated to TTL circuitry, so that the processor can handle the higher-level matter of goals and priorities.

CIRCUIT DESCRIPTION

Figure 11-1 gives the schematic for the entire Audigen subsystem. The circuit appears overly complex at first glance, but this is not so in light of what it does. The Audigen provides a complex combination of sound patterns and variations, providing the pet robot with a wide variety of expressions.

The upper four bits of the 6-bit port BC are dedicated to the controlling of the Audigen. These four bits are fed into a 7475 4-bit latch. A 7425 4-input NOR gate monitors the four bits. As long as the input is 0000, the 7425 output remains at a logical "1". But if the RCU-85 writes any other number out to the circuit, the 7425 output drops to a "0". This triggers the 74123 one-shot, which pulses the 7475 and latches the new input number. To put it plainly, the 7475 latch will latch any input as long as it is preceded by a 0000. The purpose for this will be explained shortly.

The 4-bit output of the 7475 is split into two sets of two bits each. Each set of two bits is decoded into four possible outputs by means of a 74139 dual 2-to-4-line decoder. That is, only one of four output lines will go to logical "0", depending on the 2-bit input.

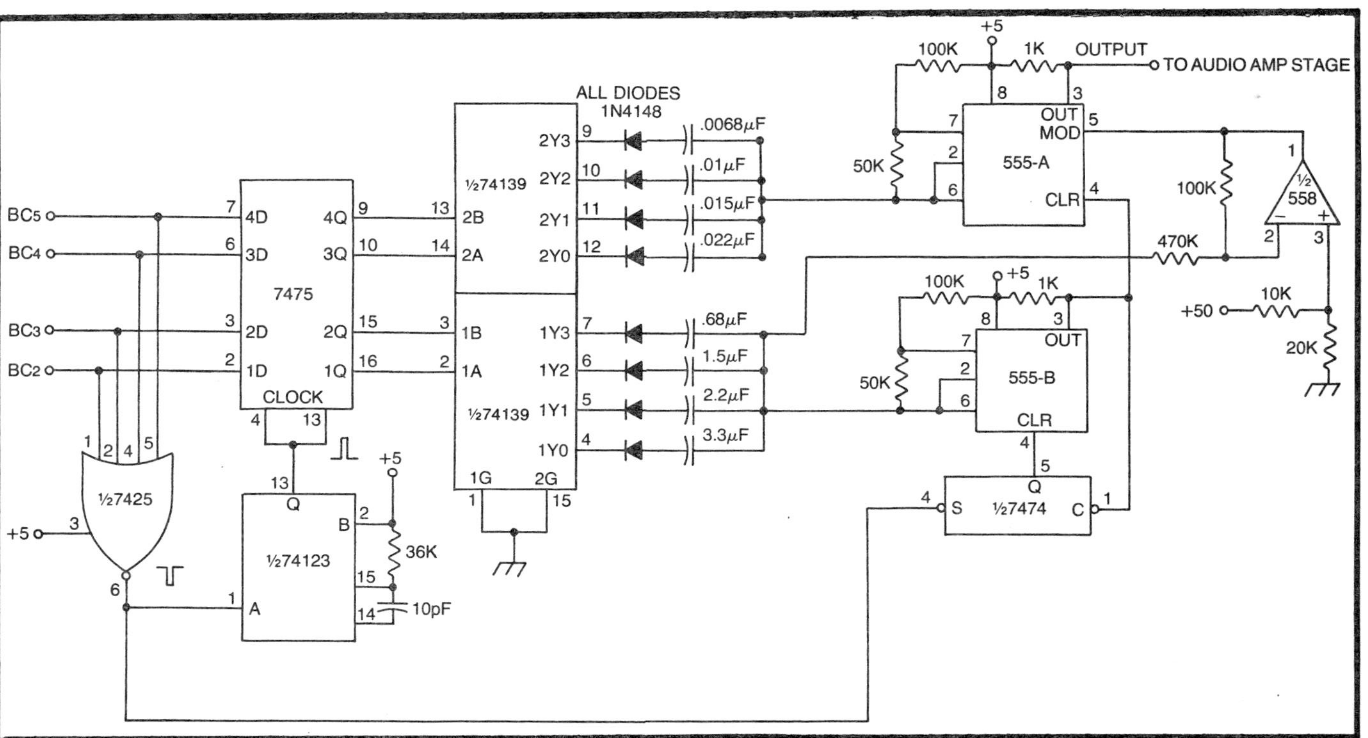

Fig. 11-1. Audigen circuitry.

The upper two bits are decoded so as to select one of four possible capacitors for the timer 555-A. This timer is wired in the astable mode, in which it generates more or less of a square wave. The frequency of the output wave depends on the four selectable capacitors. The values shown give frequencies of about 330, 480, 720, and 1060 Hz, though it is really up to the designer to tailor-make his pet's "voice."

The lower two bits are decoded to select one of four capacitors for the timer 555-B, which is also wired astably to produce a square wave. The output of 555-B directly controls the $\overline{\text{RESET}}$ pin, pin 4, of 555-A. Thus, as long as 555-B puts out a "0", 555-A will produce no output frequency. But if 555-B puts out a square wave, the resulting output of 555-A will be a "chopped" frequency, with a repetition rate directly dependent on the other timer. The capacitors shown provide rates of about 2, 3, 5, and 10 Hz, but again, the experimenter has final say.

Figure 11-2 shows some of the relevant waveforms of the Audigen circuit. The running of 555-B depends on the charging and discharging of the selected capacitor. While the capacitor is charging, the output is "1", permitting the other 555 to oscillate; while the capacitor is discharging, the output is "0" and the other 555 is silenced. 555-B itself is held at a "0" when its own RESET line is held low.

The 555-B $\overline{\text{RESET}}$ line is controlled by the 7474 flip-flop. Normally, the Q output is at "0", holding 555-B at "0", which in turn blanks out 555-A. The Audigen subsystem is thus normally silent. But suppose the RCU-85 changes the input word from 0000 to some number, then back to 0000. The 7425 output would momentarily pulse to a "0", latching the input number, and also setting the flip-flop output Q to a "1". Both 555s begin their clocking, using their selected capacitors.

The 555-B output is a "1", permitting the other timer to run while the capacitor charges. But then the 555-B output drops to a "0", blanking the other timer. The output also *clears* the flip-flop output Q, and the whole circuit returns once and for all to a silent state. Thus, if the RCU-85 issues an Audigen command and then returns the word to 0000, the subsystem will emit one "beep"—the frequency depending on the upper two bits, and the length depending on the lower two bits.

But what if the RCU-85 sends out an Audigen command and leaves it, without immediately clearing it to 0000? Then the 7425 output is fixed at "0". And so, after the first "beep," when the 555-B output tries to reset the flip-flop, it will be unsuccessful—because

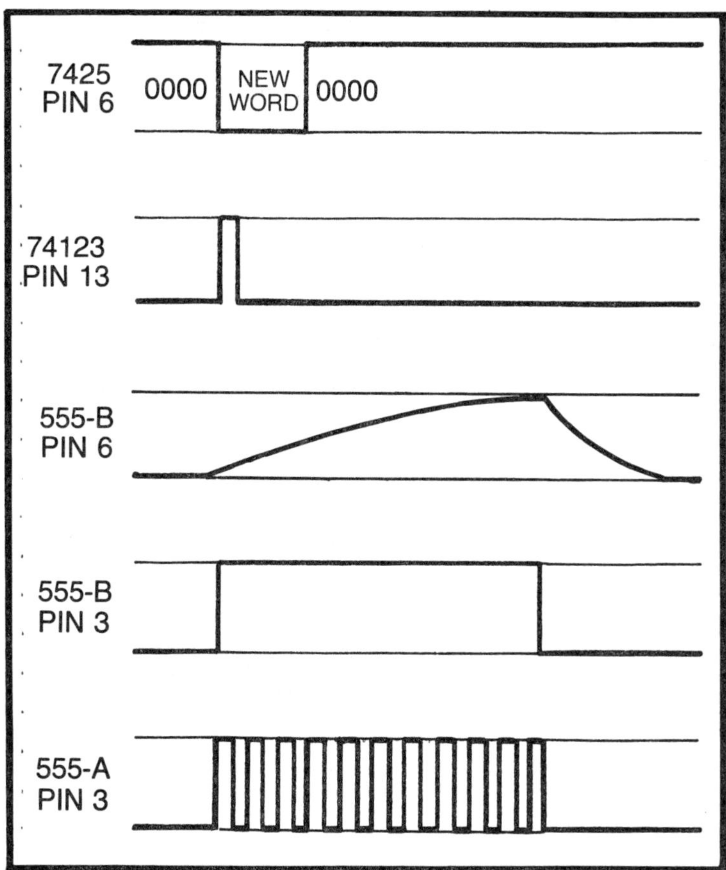

Fig. 11-2. Major waveforms for the Audigen circuitry.

the *set* input of the 7474 *overrides* the *clear* input. And so the 555-B will begin a second sequence, and a third, producing beeps as long as the RCU-85 does not clear the input to 0000. The repetition rate of the "beeps" is determined by the lower two bits.

And so, the RCU-85 has a multitude of options. It can produce single "barks" at will, by pulsing the command word; the single "bark" can have one of four pitches and one of four lengths. Or the RCU-85 can produce repeated barking, by leaving the command word valid for a few seconds; the pitch has four possibilities, and the repetition rate has four. The RCU-85 can even change pitches and rates *during* a chain of barks, as long as it makes a brief return to 0000 to latch the new information.

One more factor in the Audigen circuit contributes to a closer imitation of animal sounds. Nothing sounds more artificial and lifeless than a straight, unmodulated beep. Living pets place certain inflections on their barks and meows, and the robot pet, ideally, should do so also. Fortunately, there is a modulation pin on the 555 chip, pin 5. The pin is normally at a level of about 3.3 volts, but if the pin is externally lowered, the astable frequency increases by a certain percentage.

The 558 op amp shown in Fig. 11-1 is designed to track the gradually-rising charge voltage in the capacitor of 555-B, and to output the inverse of this curve as a modulation input to 555-A. Thus, each time 555-B is permitting 555-A to oscillate, the op amp modulates the oscillation frequency upward. The result is that each bark has a characteristic rise in pitch, sounding more like an animal "yelp" than an electronic beep. The degree of upward modulation depends on the gain of the op amp, determined by the components selected for it. In fact, it is certainly an option to configure the 558 as a *non*-inverting op amp—which would result in barks with a *drop* in pitch, rather than a rise. The technique applied here, or the use of the op amp altogether, is a matter of aesthetic appeal, and is up to the individual experimenter.

Let's review a bit what goes on. The Audigen receives a 4-bit word from the RCU-85, which it decodes to select frequencies for two 555 astable timers. 555-A generates a frequency, which is chopped by 555-B into individual barks. Depending on how long the RCU-85 leaves the 4-bit word on the port pins, the Audigen will generate either one bark of a variable length and pitch, or a series of barks of variable pitch and repetition rate. Finally, the 558 op amp provides a bit of intonation to the output of the circuit.

The Audigen circuit output is fed either into an integrated circuit or discrete transistor audio amplifier, which in turn drives a small 8-ohm speaker mounted somewhere on the sensorframe. The specific amplifier used is not crucial, but it is important to utilize one that will operate at +5 volts, which is fairly low for audio applications. The LM 386 or an equivalent circuit would fill the bill nicely, as illustrated in Fig. 11-3.

CIRCUIT ASSEMBLY

The Audigen subassembly is wired onto the brain board, with the exception of the audio amplifier, which is mounted on a small card with the output speaker. The output of the main circuit is routed to the amplifier through connector "I," pin 20.

About eleven 16-pin DIP sockets are needed for the Audigen assembly: one each for the 7475, 7425, 74123, 74139, and 7474; one each for the two 555s and the 558, with room for components; and

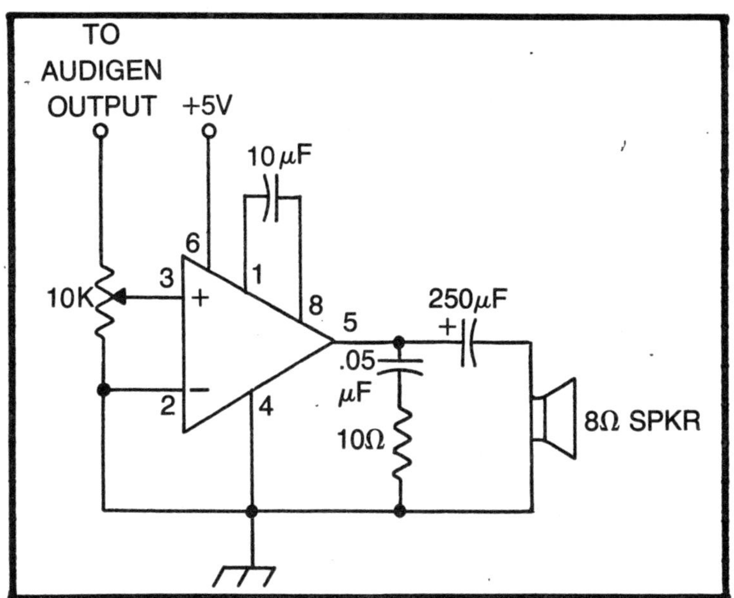

Fig. 11-3. Audio amplifier for the Audigen system.

about three more sockets for components, including the 555 capacitors, diodes, and parts for the 74123 one-shot. Of course, the capacitor values chosen, if high, may make insertion into sockets impractical due to size. In such a case, these can be wired onto the brain board itself. The diodes are necessary to prevent the unselected capacitors from affecting the circuit, since the outputs of the 74139 will pull these other capacitors up to +5 volts in lieu of the diodes.

The 74123 one-shot is wired to output a latching pulse of about 100 nanoseconds. The pulse is short, so that it can latch an Audigen command word which the RCU-85 issues and then immediately resets to 0000. The RCU-85 is likely to take as long as 10 microseconds to reset the command, and so the 100-nanosecond pulse is quick enough.

The 555 components can be reselected for different pitches, if desired. The standard formula for 555 astable frequency is as follows:

$$f = \frac{1.44}{(R_A + 2R_B)\,C}$$

where f is in kilohertz, C is in microfarads, and R is in kilohms. R_A represents the resistor from pin 8 to pin 7, and R_B is the resistor from pin 7 to pins 6 and 2.

Since the RCU-85 is not programmed and running, some source of 4-bit Audigen words will be needed. The best approach is shown in Fig. 11-4; four switches are used to ground pin locations to a logical "0," or permit them to rise to a logical "1". The flexible leads from these switches can be slipped into pin locations 39, 1, 2, and 5 of the empty 8155-B socket, thus controlling the Audigen circuit.

As long as the switches are set to 0000, the Audigen subsystem should be silent. Any other number, though, results in a series of barks. A quick *strobed* input—i.e., an input quickly restored to 0000—should produce only one bark.

If the circuit gives no output at all when commanded by the switchbank, check the 7425 and 74123; are they properly latching information into the 7475? Remember that the switches must be flipped as simultaneously as possible, since the 74123 will immediately latch the *first* change it notices. If desired, for testing purposes, the line from pin 6 of the 7425 to pin 1 of the 74123 can be broken temporarily. Then a pushbutton can pulse the 74123 pin 1 to ground *after* the four switches have been set to the desired test command.

The 558 op amp will affect the circuit drastically if it is not wired carefully. If its output is virtually at zero volts, 555-A may either be oscillating at ultrasonic speeds or halted altogether. The gain of the op amp, then, should be restricted. The resistor from pin 1 to pin 2 sets the gain; the value in the figure selects a gain of ⅓. The experimenter should substitute various gains to obtain a euphonious

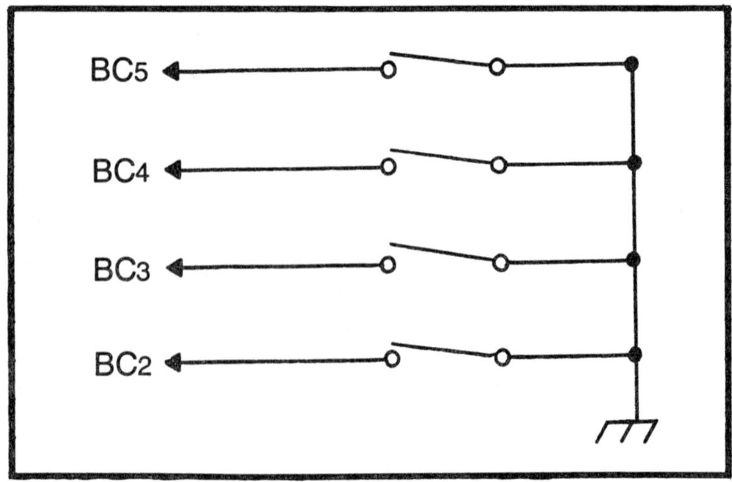

Fig. 11-4. Test input switchbank for the Audigen system.

result. The resistor divider at pin 3 may also be adjusted for a varied response.

SUMMARY

With the addition of the Audigen circuit, the major TTL peripheral circuits of the brain board have been completed. Looking back over the past few chapters, the pet robot has been given a method of controlling its own motions, of scanning its environment, of receiving vocal commands, and of producing audible responses. In most cases, the design philosophy has been to make the TTL circuits as "intelligent" as possible, so that the RCU-85 might be unburdened, free to act as overseer and goal-setter.

A few other circuits of secondary order of priority further expand the sensory capabilities of the pet robot. These possible additions will be treated next.

Chapter 12
...And Other Circuits

There is, it seems, yet another characteristic of living things that is also true of the pet robot, and that is the fact of *growth*. Life that no longer undergoes change can hardly be termed life; correspondingly, if the pet robot project is ever termed "completed," it is stagnant.

This being the case, the robot pet has purposely been designed to be expandable, capable of being the "rolling test bench" for countless new experiments. This is especially true because of the modularity of the RCU-85 control scheme. With as many as nine possible I/O ports, each of which are software-programmable for either input or output, the RCU-85 has plenty of room for enhanced sensory input and increased output expression.

In this chapter, then, let's discuss some possible add-on circuits to our system. Some of these will be so useful that you might wish to add them immediately. Others may be seeds for later thought and experimentation.

BATTERY MONITOR

The first suggested circuit is a simple convenience which will prove itself indispensible during prolonged robot pet activity. Presently, the only way to handle "feeding time" for the pet—that is, recharging—is to stop it, check its battery with a meter, and then either perform or postpone charging. Living things, though, have a somewhat better approach: a built-in early warning system known as "hunger." The family hound is usually all too aware of its need for "recharging." He needs no dinner bell to tell him!

So, in Fig. 12-1, we have the *battery monitor* circuit, a simple add-on to increase the pet's awareness of its internal environment. The key component in this circuit is the ubiquitous LM 311 voltage comparator. One leg of the comparator is held at a fixed reference voltage, determined by the zener voltage. A 2.8-volt reference serves well for this purpose. The other leg of the comparator monitors a resistor divider placed across the unregulated 6-volt battery supply. The potentiometer allows this monitor input to be calibrated for a specifically desired comparison.

In practice, the battery monitor functions in this manner. When the battery supply is at full charge, it will have a level of perhaps 6.5 volts. The battery supply drops gradually with time, until it reaches about 5.0 volts. At that point, the supply is in serious need of recharge; in addition, the output of the minimal input regulator (see Chapter 5) with this input is bordering on dropping below the specified tolerance for the microprocessor chips. Now is the time for the monitor to act.

The user calibrates the Monitor such that the output of the resistor divider equals the 2.8-volt reference when the battery voltage is getting too low—say, about 5.2 volts. Thus, as long as the battery supply is charged well, the comparator will be in a given state; but at 5.2 volts, the comparator switches state. The change is expressed by the built-in transistor of the 311, the collector of which is pulled up to +5 volts by a 5000-ohm resistor. As long as the charge is high, the transistor is on, and the output is a logical "0". When the

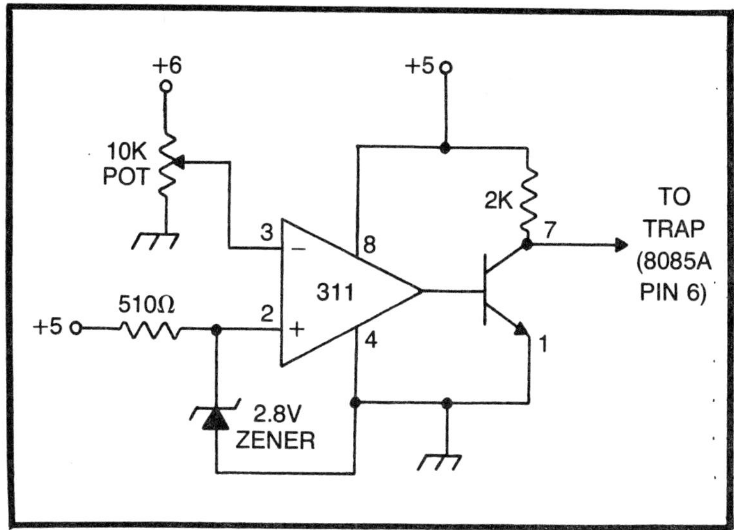

Fig. 12-1. Battery monitor circuit.

critical point is reached, the collector opens and is pulled to a logical "1".

This basic interface from the analog world of voltages to the digital realm of the microprocessor is all that the RCU-85 has been awaiting. The monitor output can now be tied to one of the special interrupt pins of the 8085A processor. As we have seen, when these four pins (TRAP, RST 7.5, RST 6.5, and RST 5.5) are activated, the RCU-85 program to jumps temporarily to a certain address, keeping track of where it was so that it can later resume the normal program flow. The highest-priority interrupt pin is TRAP; it overrides *any* program, normal or interrupt-initiated. When the TRAP pin is held at a logical "1", the program will jump to address number 36 in memory and will follow instructions from there until told to return by the program.

Immediate action! This is precisely what we need for the robot pet with regard to failing supply voltage. The monitor circuit triggers the TRAP input, and a special program is waiting, starting at decimal address 36. The special program will be described in Chapter 16, but two basic responses are involved: alarm and shutdown. Upon reaching the critical voltage level, the pet attempts to warn its owner, by a series of barks from the Audigen. If the pet is not reached by the owner in a short while, it shuts down altogether. The shutdown is accomplished by self-deprogramming of the 8155 ports and the execution of a halt instruction. Thus, the robot pet ceases all motion and thought. If the voltage of the regulated +5-volt supply should drop very low, any resulting erratic data flow is quite unlikely to reactivate the pet into haphazard action.

The battery monitor circuit takes up only two 16-pin DIP sockets on the brain board—one for the 311 comparator and one for parts—and the trimmer potentiometer can be mounted nearby. The unregulated 6-volt level is available to the brain board at pin 2 of connector "I". For calibration purposes, though, it is best to hook the resistor divider across an external variable DC supply and adjust the supply to about 5.2 volts. Then the trimmer potentiometer can be calibrated such that the two legs of the 311 are equal with such an input. If an external supply is not available, you can temporarily hook the divider across the regulated +5-volt supply and adjust the trimmer so that the monitor leg is a bit *below* the 2.8-volt reference. Then switch it over to the 6-volt battery supply, and it should be close enough to the right setting.

With the addition of the battery monitor, the robot pet has gained a *sense of hunger*, a sense which will be in its best interest and safety. In a manner analogous to involuntary reflex responses which often protect living things, the robot pet's physical "exhaustion" can now reach a point where it seizes control from the RCU-85 for "survival's" sake.

EVENT TIMER

In discussing the Excom subsystem in Chapter 10, we noted that the robot pet would expect the components of a word in "Fredian" to occur within certain time limitations. In order for the pet to make this sort of decision, it needs a general-purpose *event timer* with a wide range of programmable durations.

The 8155 timer/counter is a wide range delay device, indeed. Based on a 14-bit initialization, the timer/counter can "clock down" in any number of increments from 1 to 2^{14}, or from 1 to 16,384. With a 10-Hz input at TMR IN, it can introduce a delay from 100 ms to 27.3 minutes; with a 1-Hz input, the range is from 1 second up to 4.55 hours! Clearly then, we will have no problem producing a delay—the question is rather, "How long or short a delay are we likely to need?"

For the Excom subsystem, a time delay resolution of less than a second is desirable. After all, Fredian words operate in the order-of-magnitude of seconds. On the other end of the scale, there may be certain ARASEM "free choice" subroutines that are designed to run continuously for periods of time in the half-hour to hour range.

All things considered, the basic 10-Hz clock provides enough of a range for most applications....from 1/10 of a second up to about half an hour. The programming for Excom in Chapter 16 will assume this reference. But if the reader desires a wider range, there is a second version available.

The timer within 8155-A is available for this event timer use. Its input pin TMR IN expects a pulse train or square wave input. The 555 timer, wired astably, will provide the 10 Hz pulsed clock; see Fig. 12-2. With such a clock, the time delay simply depends on the 14-bit count length number entered into TMR LO and TMR HI; review Fig. 7-6. In this application as an event timer, the best mode would be "10," in which the count does *not* repeat, but simply results in an output pulse at the reaching of *count zero,* testable at TMR OUT.

The output ($\overline{\text{TMR OUT}}$) can now be tied directly to an input port pin, to make the timer status monitorable by the RCU-85. The prototype uses port pin BA_2 of 8155-B. In action, the RCU-85 loads TMR HI and TMR LO of 8155-A with a number representing a delay; it starts the timer by writing the correct word into the C/S register for 8155-A; then it monitors port pin BA_2, watching for a

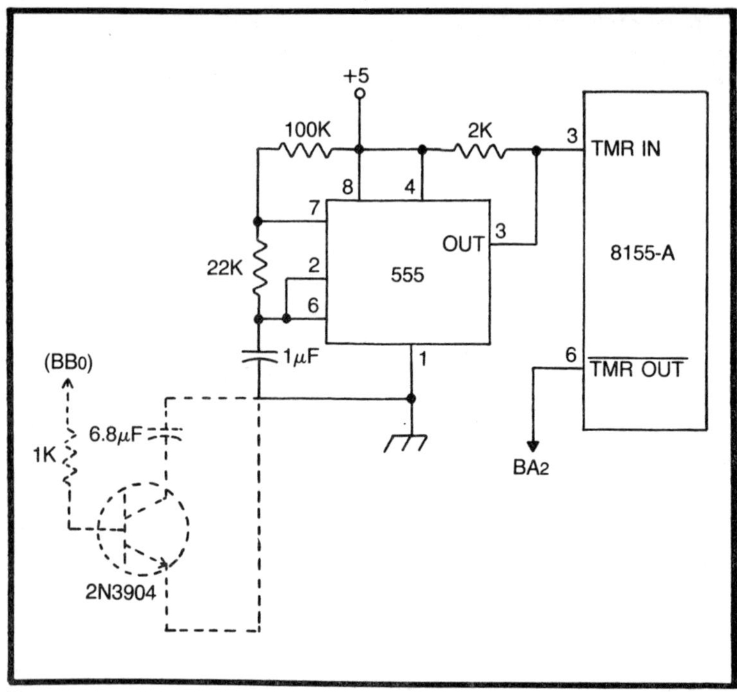

Fig. 12-2. Event timer circuit; optional range-extension circuit shown in dashed lines.

"1"-to-"0" transition. A zero at BA_2 breaks the loop, and the RCU-85 continues its program.

What if you want a greater time range? What if the reader has a special program in which the robot pet must sit around and wait for 3 hours? In this case, it is possible to use the second 555 clock shown in Fig. 12-2. This version has two selectable capacitor sizes, one resulting in a 10-Hz clock, one in a 1.3-Hz clock. The logic is wired such that a control input of "zero" selects the normal 1-Hz clock, but a"one" selects the extended version. Thus, this version is compatible with the first when the input is held at zero. A free output port pin, such as BB_0 (when BB is programmed to be an output port), is tied to the *control* input. The timer is in the normal mode if this extra port is ignored. But the extension is at least available if necessary.

Either version is simple to add to the brain board. The first version requires only one DIP socket for the 555 and the few components. The second version requires the additional capacitor and the driver transistor, which can be mounted nearby.

It is said that animals have a keen sense of time. If so, the robot pet has now gained another point in favor of life simulation. With its precise timing instinct, it can coordinate its activities according to stringent, preprogrammed goals of sequence.

CONTACT SWITCHES

As Chapter 9 claimed regarding Soniscan, a robot is better off with a distance-sensing system than simply a contact-based guidance system. But why must it be "either-or"? There are certainly cases in which the Soniscan "beam" may miss some low-lying obstacle. If so, then a tactile sensory system has a rightful place in the pet's entourage of inputs.

Switch contact inputs are a breeze for the RCU-85 to handle. It is simply a matter of wiring one side of a SPST switch to ground, and the other side to an input port pin, tied up to the +5-volt supply with a 5000-ohm resistor. As long as the switch is open, the RCU-85 sees a logical "1". Any physical contact resulting in the closing of the switch, though, shows up as a "0" level at the port pin. Figure 12-3 illustrates the idea.

Input port AB has been set aside for contact switch inputs. Up to eight switches can be handled in the manner described above. The part requiring the most consideration, though, is the physical placement of the switches. For one thing, the robot pet cannot see the floor below it, so it's particularly susceptible to falling into holes or

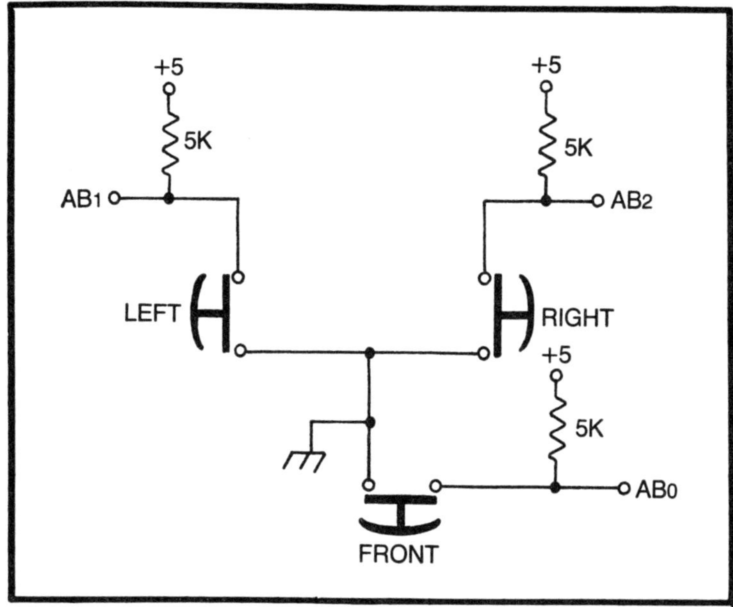

Fig. 12-3. General hook-up of bumper-contact switches.

down staircases. A good place for a contact switch, then, might be right out in front of the main drive column. An aluminum rod, ending in a small wheel of some sort, could be mounted on the drive column in front of the wheel by a few inches. The rod would be spring-loaded and coupled to a microswitch, such that the floor would normally keep the switch closed. A sudden drop in terrain would result in a change in the port pin, and the program of the RCU-85 could halt the pet instantly. How practical this suggestion is depends heavily on how quickly the drive wheel stops when instructed to do so.

We'll not go into details on the construction of uniform contact bumpers and such; but the placing of such microswitch-coupled bumpers around the perimeter of the pet would be a satisfying addition. Bumpers on the left, right, front and rear are logical; a top sensor of some kind, mountable on a protective bracket over the brain board, would help guard against damaging contacts to this sensitive area.

VARIOUS FINE TOUCHES

Pets not only express their emotions by means of barks and growls; they make use of their appendages. Wagging tails and perking ears are legitimate means of communication for the average animal. The addition of analogous equipment to the robot pet's frame may seem frivolous, but it is nevertheless consistent with our design goal of simulated life.

Figure 12-4 shows how a small gearmotor, equipped with a camshaft output, can convert rotary motion into lateral action for a wagging tail. Something of the same sort can be used to move a pair of "ears" built into the headpiece of the robot pet.

Of course, these sort of additions are much more convincing if an outer "skin" is provided for the pet. Most fabric stores will have a wide selection of shaggy fur material which can be cut and formed to the pet's body to make a removable hide. Areas of difficulty will be the ultrasonic transducers, which can only be covered with a "sound-transparent" material such as netting; and the wheels, which will require some sort of fenders to prevent contact with the material.

MOVING ON

At this stage, the reader reaches a turning-point in the construction of the robot pet. Chapters 1 through 6 majored on the physical, mechanical construction of the pet and Chapters 7 through 12 have dealt with the electronic development of the brain board. Chapters 13 through 17 will handle the final stage of the project: the

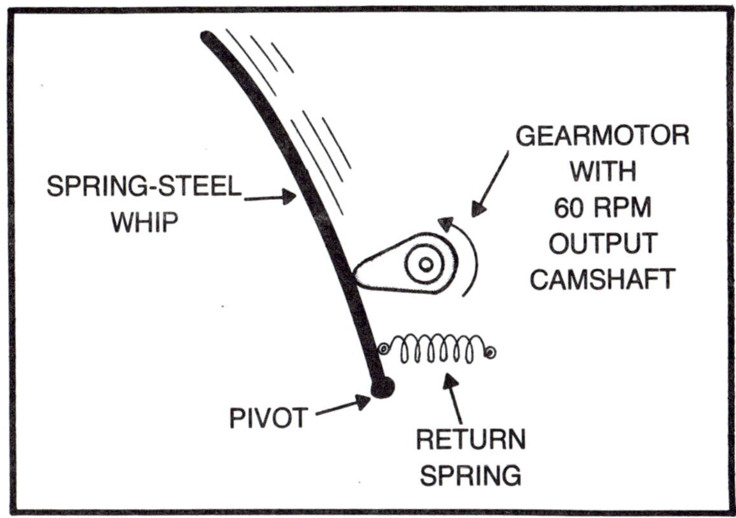

Fig. 12-4. Converting rotary motion into lateral motion for an optional wagging tail.

programming for the RCU-85. Without this last step, the pet will remain "lifeless"; only the software can effectively unite the varied subsystems of the robot pet into a cohesive whole. The program is the pesonality of the pet.

Such a mammoth advance should be taken gradually and carefully. In the next chapter, we will assemble the necessary circuitry to permit programming of the RCU-85, and the doorway to an active, lifelike robot pet will begin to swing open.

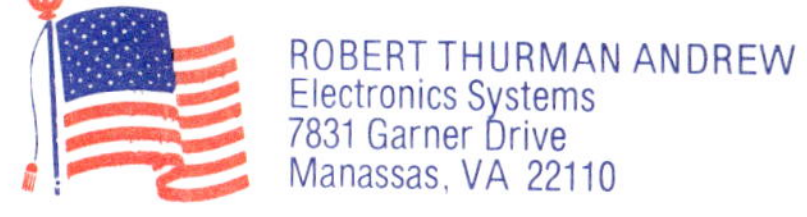

Chapter 13
Manual Programmer

The key to the lifelike performance of the robot pet is in its *programming*. It can move about, making logical decisions, provided it is given an effective program. But before we can leap into a full discussion of software techniques, we'll need a way to load such information into the inner recesses of the RCU-85.

In Chapter 4, an access cable was run from connector "C" to the low-mounted connector "A." This arrangement gives us free access to the bus system of the RCU-85 on the brain board, without having to cut into existing wiring.

The Reset Switch which has been mounted next to connector "A" also plays a role in programming. Ordinarily, while the RCU-85 is up and running, the AD lines are busily exchanging data. The state of the lines is in constant flux, undulating between TTL levels of zero and +5 volts, as binary bits pass from memory to processor, from processor to port, and so on. In short, *some* chip is always using the lines, driving them to this or that TTL level in a highly-synchronized, difficult-to-interpret fashion.

When the *reset switch* is activated, though, the whole story changes. Counters revert to zero; processes halt. In this "twilight zone" of the RCU-85, all AD lines and control lines are in a high-impedance state. They are "tristated"—that is, they are neither 1 nor 0, neither +5 volts nor zero volts. The lines are now fair game for some external device to drive them to a 1-level or 0-level; they will offer no resistance to the attempt.

This is precisely the method by which the owner can feed his programs into the memory of the RCU-85. While the system is reset, he can simply place each successive 8-bit word of his program onto the AD lines, using connector "A," and shift the proper control lines. The 8155 RAM chips have no choice but to swallow the programs into RAM. In fact, the chips would not know the difference between the human programmer and the processor. All they care about is the proper sequence of control signals.

What lines, what pins of connector "A," are the crucial ones to affect? First of all, naturally, are the AD lines. A circuit must be devised to place both data and addresses on these lines at the proper time. Then to latch the address information into RAM, we'll need to pulse the ALE line while holding the IO/$\overline{M}$ line at ground potential. Next, to latch the data information into RAM, at the locations previously addressed, we'll need to pulse the $\overline{WR}$ line while IO/$\overline{M}$ is grounded. Finally, to select the specific 8155 peripheral IC for programming, we'll need to control lines A_{14} and $\overline{A}_{15}$.

CIRCUIT DESCRIPTION

Figure 13-1 shows the entire schematic for the manual programmer which results from the above described design requirements. The marginal numbers along the extreme left and bottom of the diagram refer to pin-designations for a DB-25P, which plugs into the mating connector "A," completing the loop. Power for the programmer is actually drawn from the robot dog's own power system via pin 1, though this is by no means necessary.

The primary section of the manual programmer is the counter section. Two 7493 binary counters are used to set up an 8-bit TTL word on the *tristated* AD lines. A 555 timer is wired in the astable mode, to provide a clock of about 3-Hz for the two counters: the 555 output clocks one and/or the other 7493, depending on the pressing of two pushbuttons, INC_H and INC_L (Increment pushbuttons). The output of the counters is also fed to two 7447 BCD-to-seven segment decoder/drivers, which display the two four-bit halves on seven-segment LED displays. In this manner, using the pushbuttons and watching the displays, the user can quickly set up any eight-bit word for loading into the RCU-85.

It should be mentioned, by the way, that this scheme can be replaced by a simple bank of eight SPST switches with resistors tying the AD lines up to +5 volts, such that the switches selectively ground an AD line. The decision to do the job with TTL circuits was based mainly on the expense of switches, and partly on the simplicity of the pushbutton approach. Both methods work, so feel free to choose either based on your own stock of parts.

The second section of the programmer is the control section. Aside from pushbuttons INC_H and INC_L already described, there are four other control switches which produce the proper control signals to move the set-up data from the counter section into the RAM locations. The first two, labeled A_{15} and A_{14}, are toggle switches which control the state of these two address lines in the RCU-85. By setting these switches at 00, 01, or 10, the programmer enables, respectively, 8155-A, 8155-B, or 8155-C. A setting of 11 is not damaging, but of course it will not select any peripheral chip, either.

The third switch of the control section is the pushbutton labeled ALE. This button pulses the ALE line of the RCU-85 from its usual zero-volt level up to a +5-volt level. By pressing this button, the RCU-85 "swallows" whatever 8-bit word has been set up by the counter section, and interprets it as an address location somewhere in RAM. Once such an address is loaded into the RAM chip, the counter can be set to some desired data word which belongs in the addressed location.

The final switch of the control section, then, is the $\overline{WR}$ pushbutton. This button takes the usually-high $\overline{WR}$ line and pulses it down to ground. When this button is pressed, the RCU-85 will read the 8-bit word offered by the counter section, and load it into the previously-addressed RAM location.

Notice that pin 4 of the connector is labeled $\overline{\text{RES IN}}$, and that it is tied to ground. This is because it is absolutely imperative that all programming be done while the RCU-85 is in the reset state. The best procedure, of course, is to flip the robot pet's reset switch, plug in the programmer and use it, remove the programmer, and then restart the pet. If the reset switch is accidently bumped during programming, however, a shorted condition will result, since both the 8085A and the programmer will be fighting for supremacy on the AD lines. And so, as a precautionary measure, the very connecting of the programmer forces the robot pet into the reset state, regardless of the status of the reset switch on the mainframe.

The manual programmer as shown in the Fig. 13-1 has no provision for *reading* the memory locations just loaded. This addition would complicate the circuit quite a bit, though it could be done. As it stands, programming and reprogramming is such a quick and simple proposition, that if there is any doubt as to whether a word has been properly loaded, the programming sequence can be repeated easily enough at the questioned address location.

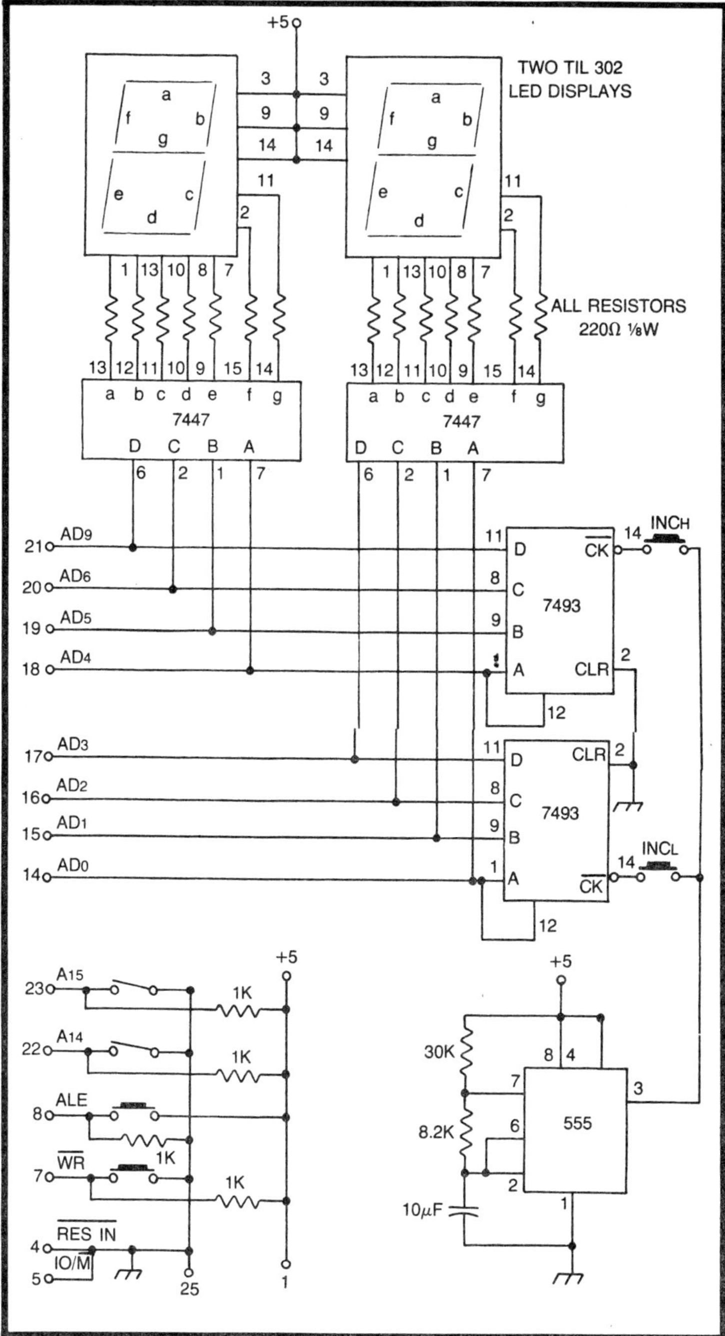

Fig. 13-1. Basic manual programmer for the RCU-85.

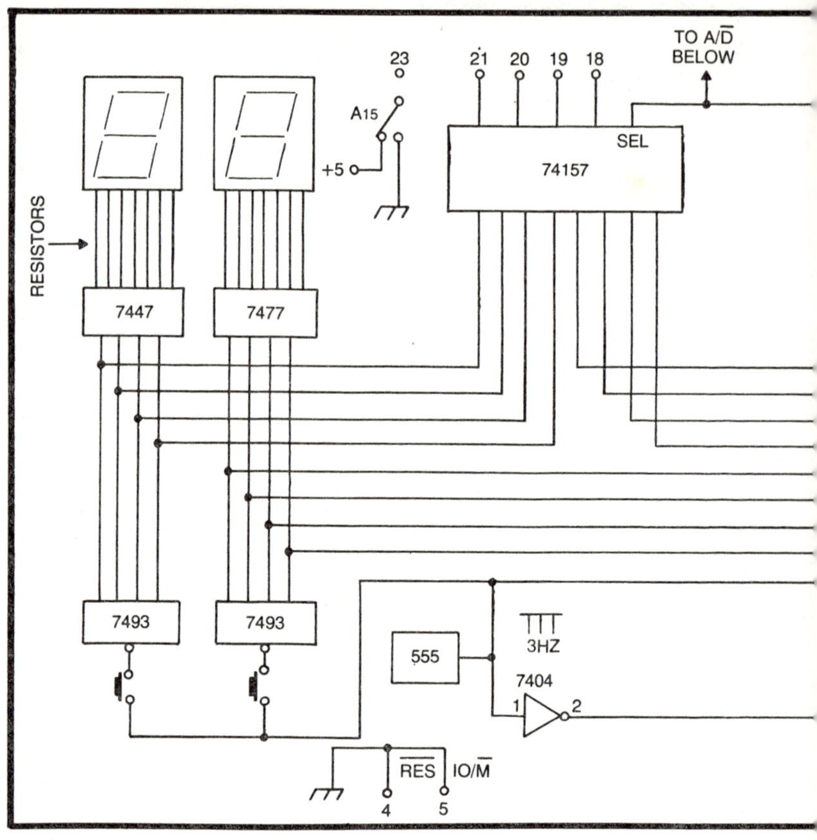

Fig. 13-2. Alternate "deluxe" manual programmer for the RCU-85.

ALTERNATE PROGRAMMER

The major disadvantage to the programmer previously described is the need to repeatedly re-establish the address information. That is, the user sets up an address, presses ALE, then sets up the data for that address and presses $\overline{WR}$; then he must load the next address and press ALE, and so on. It would be better if there were *two* counter sections, one for addresses and one for data. That way, the address could simply be incremented by one at the touch of a buttom after each data loading, thus minimizing confusion and errors.

Figure 13-2, then, gives the general idea of what such a *deluxe* manual programmer might look like. (For simplicity, not all pin designations have been shown; missing numbers can be figured out by comparison to Fig. 13-1.) There are two counter sections in this

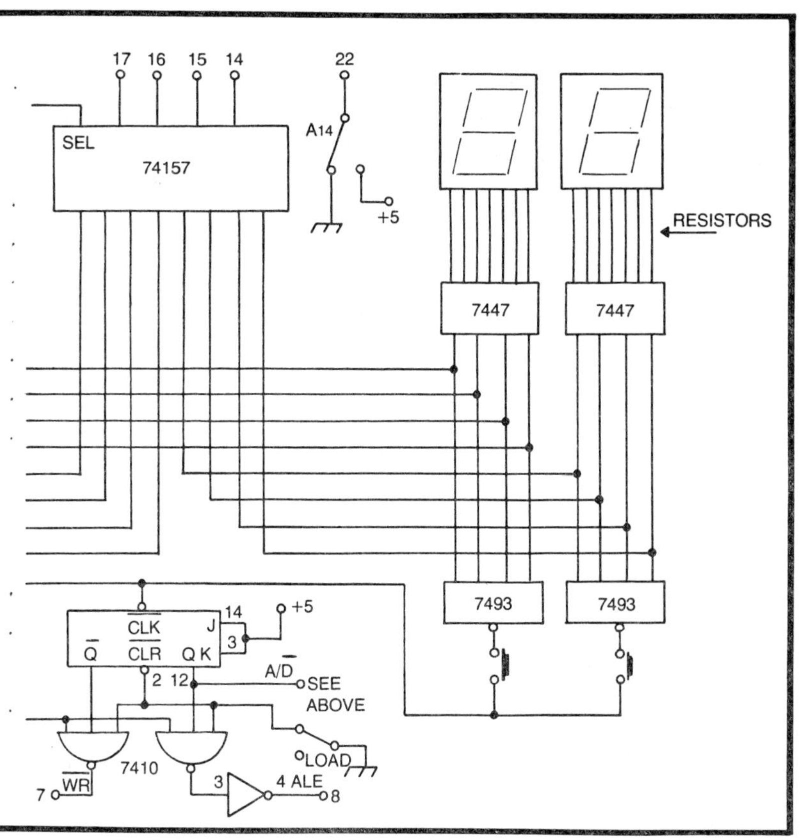

version, for addressing and data. To keep them separated in terms of using the AD lines, a pair of 74157 data selector chips are used. These two chips read both counters, but only permit one 8-bit word to reach the AD lines, depending on the state of the SEL pins. Presiding over the entire system is a 555 (the same one that clocks the counters), which produces the pulses for the ALE and $\overline{WR}$ lines of the RCU-85 and also controls the data selector chips. In practice, the user sets up his address and data words, and flips the LOAD switch. After two pulses of the 555—i.e., in somewhat less than a second—the data has been loaded into the proper address location.

Which version, the *basic* or the *deluxe*, is the best to build? Certainly the fancier second version makes things a bit easier in the end, but you may not wish to spend very much time on a "perfect" programmer while your main project—the robot pet—sits patiently waiting. I have personally found that the simpler version, once it is used for a time, is not very hard to master. In fact, the need to

Fig. 13-3. The completed basic programmer. Note the DB-25, intended to plug into the robot pet's Connector A.

repetitively re-establish the address forces the user to double-check his program lisiting, which helps prevent careless mistakes. And the simpler version, in the final analysis, consists of less than half of the number of chips needed in the *deluxe* programmer. In the end, though, the reader decides.

CONSTRUCTION

Either programmer is wired onto a piece of perforated board using the same DIP-socket approach that the brain board used. The *basic* version requires nine 16-pin DIP sockets: one for the 555 and its components; two for the 7493 counters; two for the 7447 decoders; two for the LED displays, and two for the driver resistors accompanying the LED readouts. The tie-up resistors associated with the control section can be mounted right at the switches. The deluxe version, on the other hand, requires some 22 DIP sockets: eight for each counter section after the analogy of the basic version; one for the 555 and its components; one each for the 7473, 7404, and 7410; and two for the 74157 data selectors.

If the reader is clever enough, he can lay the sockets out on a board in such a way that the assembly can be mounted into a metal box or housing, with the LED readouts visible through a plastic window. Figure 13-3 shows how the completed basic programmer appears using this technique. The goal is to make the programmer as rugged as possible, so that it will not wear out from continual use.

The currents involved in the process of programming are low. So the DB-25P of either version can be hooked up using a light-gauge insulated wire, such as 26-gauge. The connector will probably be the victim of the most abuse, so avoid using single-strand wire that tends to break after the second and third bend.

FINAL TESTING

When either version is powered up, the LED readouts should light up at some random number. The pressing of the associated *increment* button should result in a sequential upward count as the button is held down. If, in use, the speed of the count is too fast to be practical, the 30 K-ohm resistor of the 555 can be changed for a larger value, such as 56 K, which would provide about a 2-Hz clock. If the count tends to skip certain numbers or act erratically, the placing of a capacitor at pin 3 of the 555 may help; .001 microfarads is a good starting choice.

Remember that the LED displays represent 4-bit numbers. Therefore, the display does not stop at "9" or "0", but it goes on to represent the binary numbers from ten to fifteen. Table 13-1 contains a handy visual conversion chart for the output of the LED displays. In text notations, the higher numbers will be referred to using the letter notation, such as A, B, C, D, E, and F. But the programmer will instead generate the unusual forms as illustrated. These forms will become "second-nature" to you as you use the programmer, so don't be overly concerned. Note especially that the symbol for the binary number "fifteen" is an absolutely blank display. Get used to this system of notation, which is termed *hexadecimal* (meaning based on sixteen), because it will simplify program-listings somewhat.

Table 13-1. Hexadecimal Conversion Chart

DECIMAL	BINARY	HEX CODE	HEX DISPLAY
0	0000	0	0
1	0001	1	1
2	0010	2	2
3	0011	3	3
4	0100	4	4
5	0101	5	5
6	0110	6	b
7	0111	7	7
8	1000	8	8
9	1001	9	9
10	1010	A	⊏
11	1011	B	⊐
12	1100	C	⊔
13	1101	D	⊆
14	1110	E	⊥
15	1111	F	(BLANK)

Next, check out the signals generated by the control section of either version of the programmer. In the basic model, check to see that the various connector pins actually change to the proper state when the switches are activated. In the case of the deluxe version, flip the LOAD switch and monitor ALE and $\overline{WR}$. As long as LOAD is flipped, there should be a positive pulse at ALE and a negative pulse (out of phase with the other) at $\overline{WR}$. When the LOAD switch is at its grounded position, however, neither line should be pulsed, and the 7473 Q output should be reset to a zero. Monitor also the AD lines, either with an oscilloscope or with the status display device constructed in Chapter 10. The AD lines should visibly alternate between the 8-bit output of the address counter and that of the data counter.

In the end, the best way to test the workability of the manual programmer is to program the robot pet. The specifics of software development and implementation for the pet are dealt with in Chapters 15 and 16. But there is one last device to construct, the helpfulness of which will far outweigh its complexity. Although the device is not absolutely demanded by the robot pet project, there are aspects of usage which are impractical without the help of this device. And so we take one last detour before software skills to deal with the construction of a *tape interface* for the robot pet.

Chapter 14
Tape Interface

We mentioned in an earlier chapter that the major drawback to the use of *random access memory* (RAM) for the robot pet's programming is that it is "volatile"—when the power is gone, so is the program. The best way to lessen this disadvantage is to provide some quick, easy method of *re*programming the robot pet. This chapter will discuss the construction and implementation of a *tape interface* for the pet.

The previous chapter described the construction of a manual programmer for the RCU-85, but can you imagine having to reprogram by hand, step by step, any time the pet misses a charge? The tape interface will enable you to save that program on a standard cassette tape for later use. And since the whole of the pet's memory is only 768 bytes long, the process of retrieval is quite rapid.

In practice, the procedure runs like this. On paper, using the techniques described in the final chapters of this volume, you develop a program. You load it into the RCU-85 via the manual programmer. If you later determine that modifications are necessary, you alter it with the manual unit, until you're satisfied. Then you plug the pet into the tape interface, which records the perfected program onto a tape. From then on, you can reload the saved program, using the interface in the reversed mode, in a matter of a minute.

DESIGN CONSIDERATIONS

Personal computing enthusiasts will once again be ready to take me to task when they see the final schematic. After all, the tape interface requires quite a few TTL ICs, and is a complex, synchronized circuit. Some will ask ,"Aren't there several single-chip tape interface ICs on the market now?" Yes, and there are full-blown tape interface devices for sale in any conceivable mode of tape storage, but the robot pet cannot use them. Why not? The primary reason is that the robot pet has no program to do so. These commercial tape interfaces assume the presence of some sort of software monitor to operate them. The pet robot, though, will not be up and running when the tape storage or retrieval occurs. The pet has *no* permanent storage for a software monitor; it is all RAM. And so, the tape interface must do *all* of the work, including generation of proper control signals for the RCU-85. It can expect no help from the pet, who is quite immobilized in the reset state.

The tape interface, then, is much like the manual programmer, in that it provides the proper levels to the ALE, $\overline{RD}$ and $\overline{WR}$, A and AD lines, in order to access the desired memory locations. But everything is handled under the careful synchronization of a TTL counter circuit, which sets up the desired data, converts it to recordable tones or decodes it from these, and clocks the various control lines. The process begins at address location "zero" and does not stop until all of the robot pet's memory has been read or written.

Now, cassette tape recorders of the less expensive variety are simply not high quality, computer grade instruments. Their frequency range is rather limited, owing to the slow tape speed and related factors. A standard low-cost recorder may have a frequency response of from 40 Hz to 10 kHz at best. For the sake of reliability, then, our data will have to be converted into audible tones capable of being handled by these limited machines. Of course, if you have a high-quality open-reel tape system, you're fortunate. But let's try to keep things simple.

Assuming we wish to convert binary "1's" and "0's" into audible tones, what frequencies should we use? The *Kansas City Standard* format uses a 1200-Hz tone for a zero, and a 2400-Hz tone for a one. The drawback to this choice is that the tones are exactly one octave apart, and tone decoding circuits can be susceptible to confusing a desired frequency with its harmonics. *Modems*, devices for data transmission on phone lines, use 1070-Hz and 1270-Hz, and often 2025-Hz and 2225-Hz tones. These choices avoid the harmonic problem but require a sensitive decode circuit to discriminate between such close frequencies. In addition, the frequencies are fairly slow, because the limitations of standard phone lines are even greater than those of cassette machines. Because of the lower frequency, less data can be packed into a given time period.

The above methods use a data encoding approach termed *frequency shift keying* (FSK), i.e., the use of two different tones to indicate the two data levels. However, it is possible to encode data using only one frequency with a technique called *pulse width modulation* (PWM). In PWM, the difference between the data states is determined by the duration of the one pitch. The "1" tone, for example, might be twice as long as the "0" tone. The PWM approach is more susceptible to noise than the FSK method, but a good quality cassette can help alleviate this problem. The chief advantage is simpler design.

CIRCUIT DESCRIPTION

The robot pet's tape interface, a PWM system, consists of three major circuits rolled into one interactive unit. The first circuit is a sequencer which steps through the RAM memory of the RCU-85, converting all data into a lengthy, synchronized string of long and short pulses of a carrier frequency, and feeding these out to a cassette recorder. The second circuit receives the output of a cassette machine and decodes the various pulses into binary bits, subsequently assembling them into 8-bit words and writing them sequentially into the RAM memory of the RCU-85. The final circuit is a display stage which enables the user to manually step through the robot pet's memory and examine the contents.

Figure 14-1 pictures the prototype version of the interface. Three cords, not seen, extend from the rear of the unit: two audio

Fig. 14-1. Robot pet tape interface in completion.

Fig. 14-2. Basic boards for the tape interface.

cables to conduct data to and from the cassette machine, and one cable terminated with a DB-25P connector designed to plug into the robot pet's connector "A." The front panel is actually a piece of tinted plastic, behind which are two seven-segment LED readouts, which display all data in hexadecimal figures. There are four basic controls. One toggle switch selects the mode, either DATA STORE or DATA RETRIEVE. A second toggle specifies whether the DATA STORE mode is conducted AUTOmatically (actual tape storage) or MANually (memory examination). One pushbutton starts the interface into its automatic cycle, which will proceed until all of RAM memory is either stored or restored. A second pushbutton is used to step manually through memory.

In final construction, the interface consists of two boards, pictured in Fig. 14-2. The larger board is the interface proper, bearing some 30 integrated circuits. The smaller board bears the two seven-segment readouts, together with their associated decoder chips and resistors. There is no power supply in the unit, since the circuit can derive its +5-volt requirements from the robot pet's own power system, via connector "A," pin 1.

The tape interface is obviously a formidable task, but a necessary one. Without it, the pet robot *must* remain powered-up continuously to retain whatever complicated program has been fed into it. Remember, also, that the whole problem is based on the use of RAM in the pet robot's RCU-85. The only other option is to use EPROM, which would require buying an EPROM eraser and buying (or building!) an EPROM programmer. The cost of the EPROM

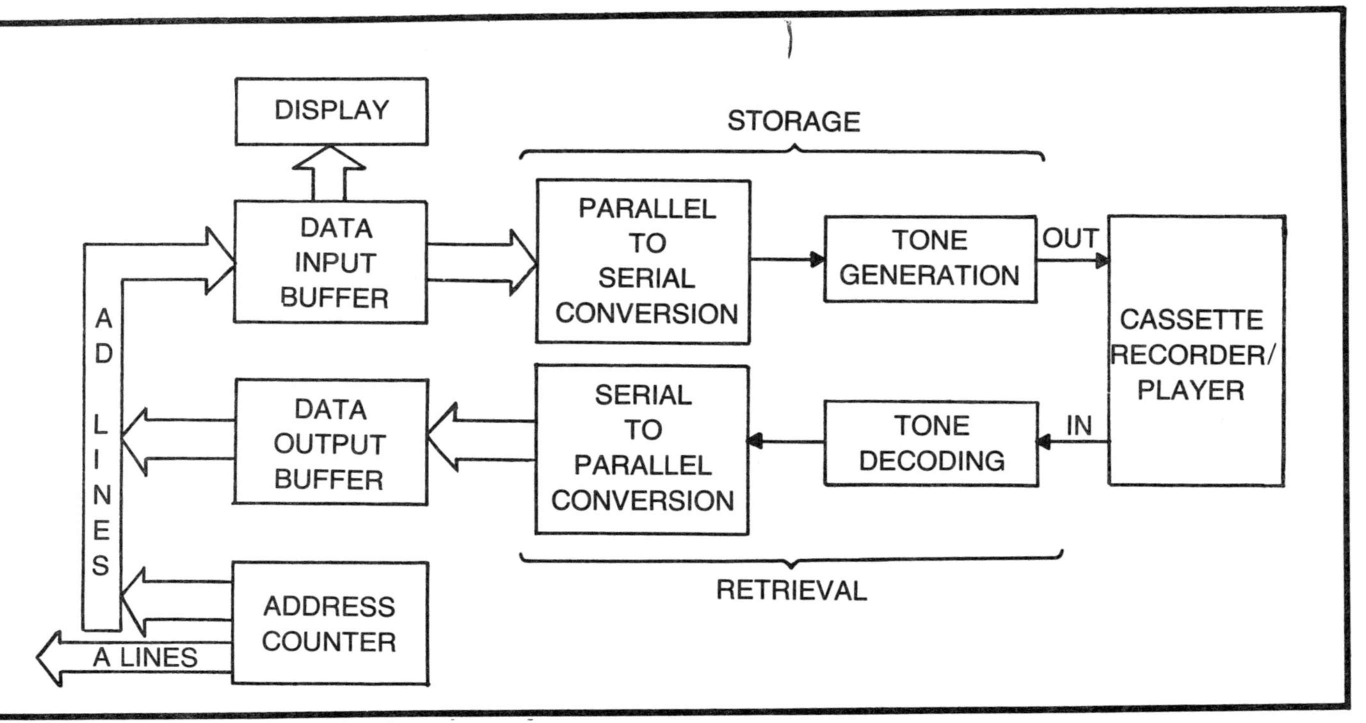

Fig. 14-3. Block diagram for the tape interface.

route sent me back to the tape interface approach, and I believe the decision is justifiable.

HOW IT OPERATES

Figure 14-3 provides a block diagram of the tape interface. The upper portion represents the *storage* function of the unit. Data from the RCU-85 is read through an input buffer, and it is simultaneously displayed. (The display is for the *manual* mode, for reading memory at leisure.) The 8-bit words are synchronized into groups of serial data pulses of varying width, beginning with the least significant bit. The pulse string is then converted into 5000-Hz tones and recorded onto tape. The *retrieval* function operates in the reverse fashion. The tones are decoded into a string of data pulses. The pulses are converted into data levels of "1" and "0", and are assembled into 8-bit words. These words are then written into the RCU-85 RAM memory through the output buffer. An *address counter* continually updates the memory address in both functions, stopping only when all of RAM has been read or written.

The following is a step-by-step description of how each of the functions is performed. It is difficult to explain one function at the exclusion of the other, since some ICs play roles in both. But the ideas should become clear with careful reading and cross-comparison between diagrams.

STORAGE FUNCTION

A schematic of those circuits relating closely to the *storage function* is given in Fig. 14-4. Circled numbers indicate pin numbers of the DB-25 output connector which mates to connector "A" on the robot pet's mainframe. In the case of NAND gates and inverters, no IC number has been noted—only a letter. Any two-input NAND gates shown belong to a 7400 IC, which holds four such gates. Any inverters shown belong to a 7404 IC, which holds six inverters. Any three-input NAND gates belong to a 7410 IC, which holds three such gates. The letter given identifies one of several ICs of the same kind—for example, a two-input NAND gate labeled "C" belongs to 7400-C. Three-input NAND gates have no letter-label, for there is only one 7410 in the entire unit.

The timing for the *storage function* is provided by 555-A, which generates a 40-kHz pulse wave. This wave is divided down by a 7493 binary counter, whose different outputs are gated to provide the major control waveforms for the RCU-85. Figure 14-5 gives these waveforms as they occur. For the majority of the time, the RCU-85 is being told to read ($\overline{RD}$) its memory into the interface. For short periods of time, though, the interface retracts the $\overline{RD}$ request, and instead issues an ALE, latching the latest RAM address of the counter into the RCU-85. The level labeled $\overline{TBF}$ will come into play when we see the counter later in Fig. 14-7. It controls a *tristate buffer* which in turn controls the flow of counter addresses out to the AD lines. When $\overline{TBF}$ is high, the buffer is *off*, preventing the counter output from shorting to the RCU-85 output. When $\overline{TBF}$ is low, the buffer permits counter addresses to be placed onto the AD lines; an ALE pulse then loads the current address into the RAM chips. In this way, the address is currently updated about 5000 times per second.

The output of 7493-A is a 5000-Hz square wave. This is fed to the output gate; it will be the tone frequency. It is also fed to pair of 7490 counters, which divide it down to about 100 Hz; this will be the bit rate, or "baud rate"—the synchronization for the serial data. To those who have worked with tape interfaces before, this may seem to be painfully slow. But remember we are dealing with less than a *K* of memory at the most. 100 Hz is plenty fast enough; the whole RCU-85 memory can be transferred in just over a minute.

The 100-Hz wave simultaneously clocks two one-shots, both of which are contained in the 74123-A package. These are timed so as to provide a short and a long pulse, representing a "0" and a "1". A new pulse is triggered every 10 milliseconds; the short pulse is 4 ms, the long is 9 ms. Further circuitry will select which pulse to send to the output gate, based on the bit being read from RAM.

Hooked to the AD lines is a 74151 data selector IC. Essentially, the 74151 will output the status of any one of the eight input bits, based on a binary input number of from 000 to 111 (zero to seven). And so there is a 7493-B counter which, starting at 000 and counting upward to 111, clocks the 74151 upward through all eight bits. Figure 14-6 shows this clocking. The 7493-B starts at 000. The 100-Hz clock triggers the two one-shot pulses, which in turn *latch* the output of the 74151 using a 7474 flip-flop. For the next several milliseconds, one or the other one-shot pulse is permitted by the gating of the flip-flop to reach the interface output stage. Before the pulse is completed, the 7493-B is clocked to 001; then the whole

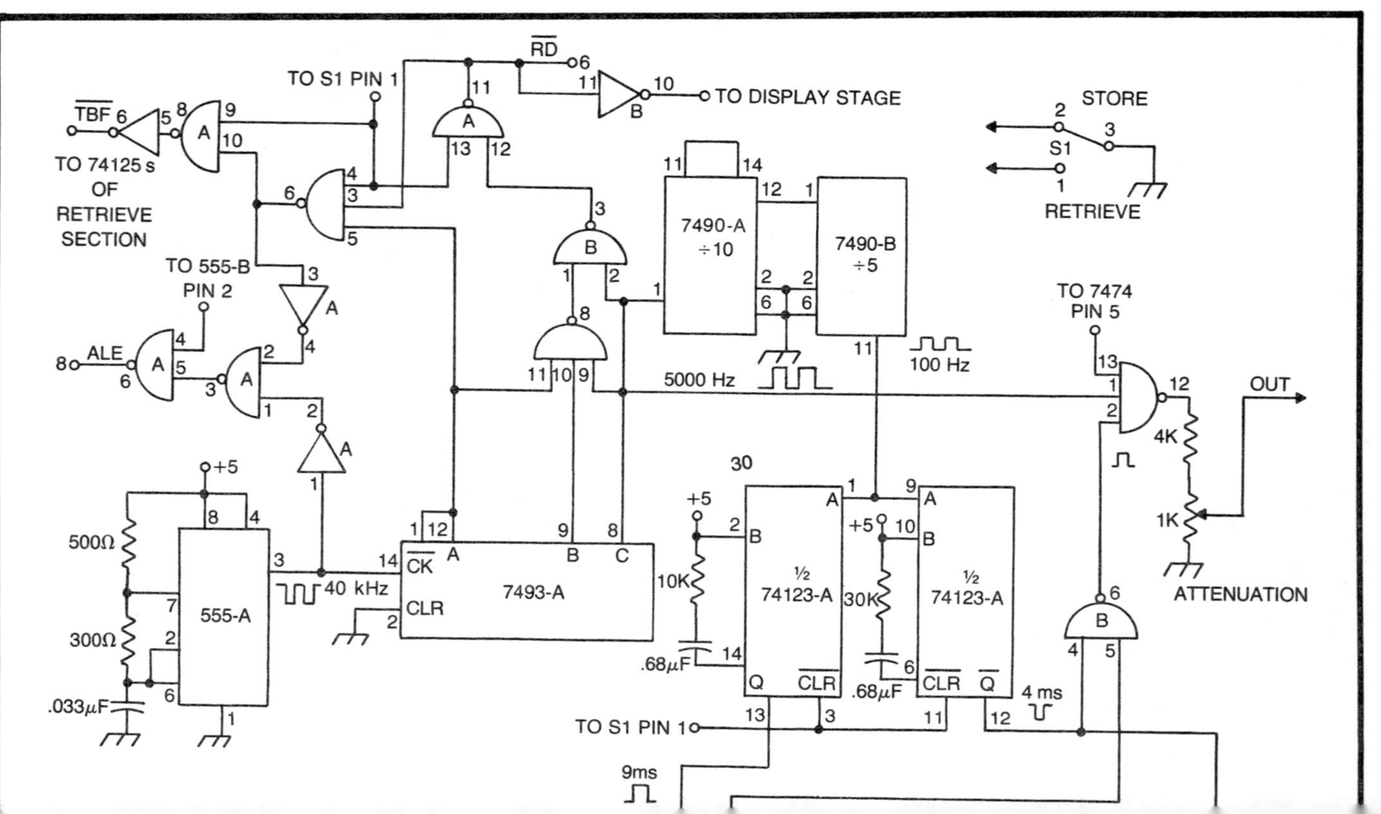

STORE
RETRIEVE
S1
TO DISPLAY STAGE
OUT
ATTENUATION
4K
1K
TO 7474 PIN 5
100 Hz
7490-B ÷5
7490-A ÷10
5000 Hz
½ 74123-A
½ 74123-A
4 ms
+5
30K
10K
.68µF
.68µF
30
9ms
TO S1 PIN 1
RD
7493-A
CK
CLR
40 kHz
555-A
TO S1 PIN 1
TBF
TO 74125 s OF RETRIEVE SECTION
TO 555-B PIN 2
ALE
500Ω
300Ω
.033µF

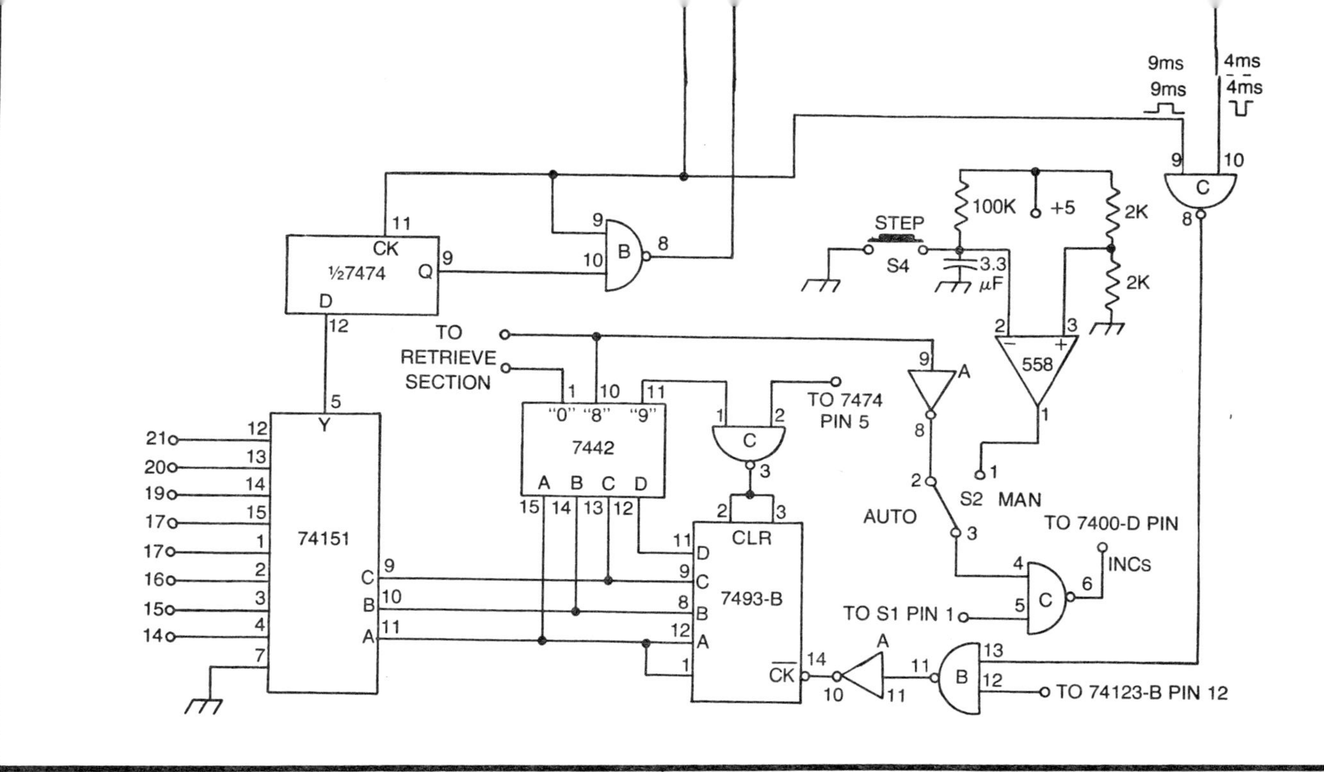

Fig. 14-4. Storage section of the tape interface.

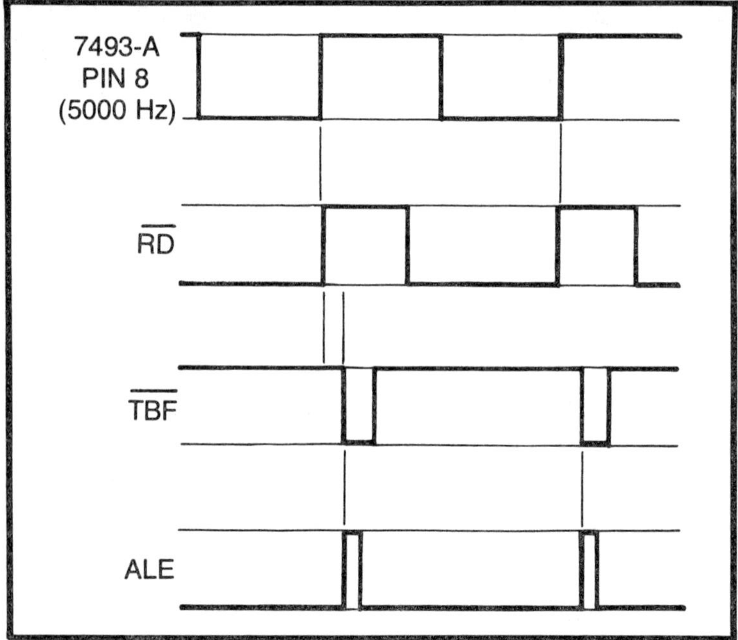

Fig. 14-5. Storage section timing for the control of the 8155 chip.

process is repeated, based on the next bit in line. This continues until all eight bits have been outputted as long or short tones.

A 7442 decoder IC is hooked to 7493-B; its numbered outputs drop to zero depending on the binary number. So when 7493-B counts through the bits up to 111 (or 0111, decimal 7), all goes as described. But when the count reaches 1000, decimal 8, the 7442 puts out a zero at its "8" line; this pulse is used to clock the address counter to the next RAM location (the address counter is pictured in Fig. 14-7). During this #8 count, the interface will put out another tone-pulse, but its state is irrelevant, and will be ignored by the later *retrieve mode*; it is merely used for synchronization purposes. Finally, the 7493-B will reach count #9, binary 1001; the 7442 uses this event to *reset* the 7493-B back to 0000 instantaneously, so that the next 9-bit word may be disassembled as before.

One more portion of the storage function circuitry remains to be explained, and that is the 558 op amp. As we have said, the 7442 pin labeled "8" ordinarily increments the address counter, so that the interface reads through memory very rapidly. But if switch S2 is flipped MANUAL instead of AUTOMATIC, this is not permitted. Instead, the address counter can be incremented slowly and manually, using pushbutton S4. The purpose is for the sequential reading

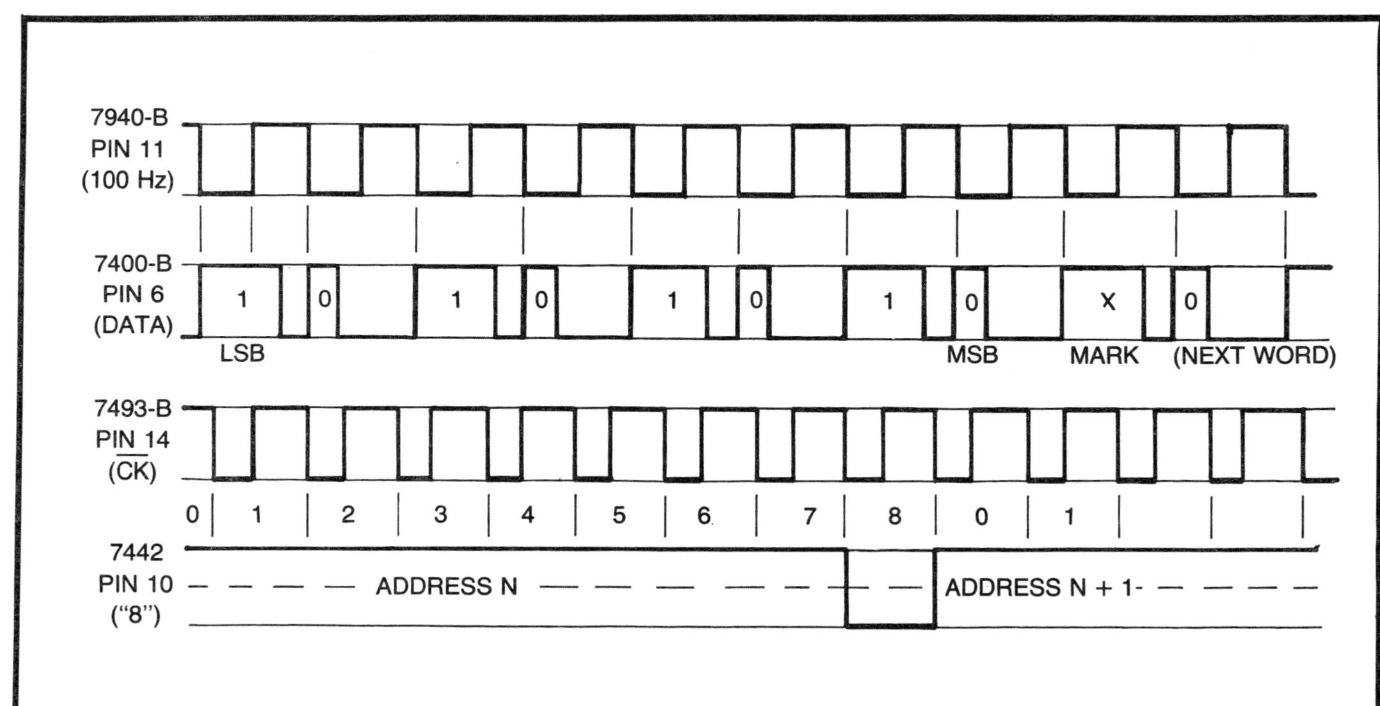

Fig. 14-6. Storage section timing for the generation of data pulses.

of RAM memory, using the *display* stage which is always monitoring the AD lines. Naturally, as this interrupts the careful synchronization of the AUTOMATIC mode, the *manual read* use should not be done while actual tape storage is occurring; it is a peripheral function of the interface. (In the AUTOMATIC mode, the display readouts are blinking so fast that they appear to be "88".) The 558 circuit is merely a debounce stage for the STEP pushbutton S4.

RETRIEVAL FUNCTION

Now let us consider the *retrieval* function with its associated circuitry. Figure 14-7 illustrates the entire set of circuits involved, except those few which play a role in both functions but are shown in Fig. 14-4. The all-important *address counter* is pictured herein, as well as the previously-described *tristate buffer*.

The retrieval process is not timed by hardware—it "self-synchronized." That is, the data pulses on tape provide the necessary synchronization for themselves. Thus, no precise clock is necessary at this stage. This simplifies the procedure somewhat.

Before we trace the inputting of a string of data tones, let's look at the layout of the output buffer stages. There are two buffer ICs, 74125-U and 74125-L; these permit data to flow onto the AD lines when $\overline{\text{TBF}}$ is low, but go to an open-collector state when $\overline{\text{TBF}}$ is high. In the storage function, these were absolutely necessary to prevent any conflicts between interface addressing output and RCU-85 RD data input. In the retrieval function, however, the interface will do *all* the talking on the AD lines; the RCU-85 will merely receive either data or addresses. And so, the *function* switch S1 in Fig. 14-4 is gated so that, in this mode, the 74125 buffers are continuously on, rather than on-and-off.

Now there is a new conflict, however, and that is between *address* information and *data* information. These two outputs are kept separate by a pair of data selector chips, 74157-U and 74157-L. Together, these chips choose to output one of two 8-bit input words, depending on the state of their SELECT pin. When SELECT is "1", the chips select the address counter; when SELECT is "0", they select the data circuitry. One or the other word is then fed out to the AD lines through the permanently-on tristate buffer stage. (In the storage mode, only addresses are needed as output information; so the SELECT pins are so gated that only the address counter is selected during the storage function.)

Study the address counter, shown in the lower half of the schematic. The counter is made up of two 7493 counters, to provide an 8-bit address for RAM, plus a 7473 dual flip-flop to provide the extra two bits for the selecting of one of the three 8155 chips in the RCU-85. (These two upper bits go directly to A_{15} and A_{14} via the output connector, not through the buffer stage.) The address counter is incremented either by the 7442, as we saw in the storage mode, or by the retrieval circuitry soon to be described. Controlling the address counter progress is a 7474 flip-flop, which acts as a *run halt* element. Ordinarily, 7474 is reset, and its *Q* output is at zero; this *Q* output is used as a RESET for various parts of the interface. But when the user presses pushbutton S3, the 7474 is set, permitting the address counter to run. The interface will continue to run until the upper two bits of the address counter equal 11—that is, when all three RAM chips are finished. At this point, the 7474 is reset, and the interface returns to a neutral state.

Let's proceed to the actual *retrieval process*. The tape player output is fed into a 567 tone decoder, which is configured to recognize a 5000-Hz tone. The output pin of the 567 will drop to zero for a long or short duration, depending on the logical value of the recorded tone. The pulse will last around 4 ms for a "0" and around 9 ms for a "1". Figure 14-8 shows this waveform and the other signals which the *interface retrieval* section generate.

The falling edge of the 567 output triggers a one-shot (one-half of 74123-B), which outputs a 100-microsecond pulse. This pulse in turn triggers a 555-B timer, set to output a pulse of about 6.5 milliseconds. The reader may already see the concept; at the instant when the 6.5-ms delay runs out, the 567 output can be checked somehow. If the 567 output is still at zero when the 555-B runs out, the pulse represents a "1"; if not, it represents a "0".

To perform this differentiation, the circuit uses a 74164 shift register. Pin 1, labeled A, is the input, and pin 8 is the CLOCK. When the 74164 is clocked, it reads whatever is at its pin A, and stores it. The next time pin 8 is clocked, the prior data is shifted down one place on the output pins, and the new data is stored. This can be done eight times, so that the eight output pins finally are outputting a reconstructed 8-bit word. The 567 output, then, is fed into pin A, and the 555-B output is gated so that, when the delay runs out, pin 8 of the 74164 is clocked. The 74164 output can then be written into the RAM location already addressed by the address counter.

The half of 74123-B which triggers the 555-B is not only for that purpose. The 100-microsecond pulse is also gated to become the ALE pulse for addresses. Each time a bit-tone is received, the ALE line is pulsed, updating the current address. This ALE pulsing occurs only during the duration of the 555-B 6.5-ms time delay. The

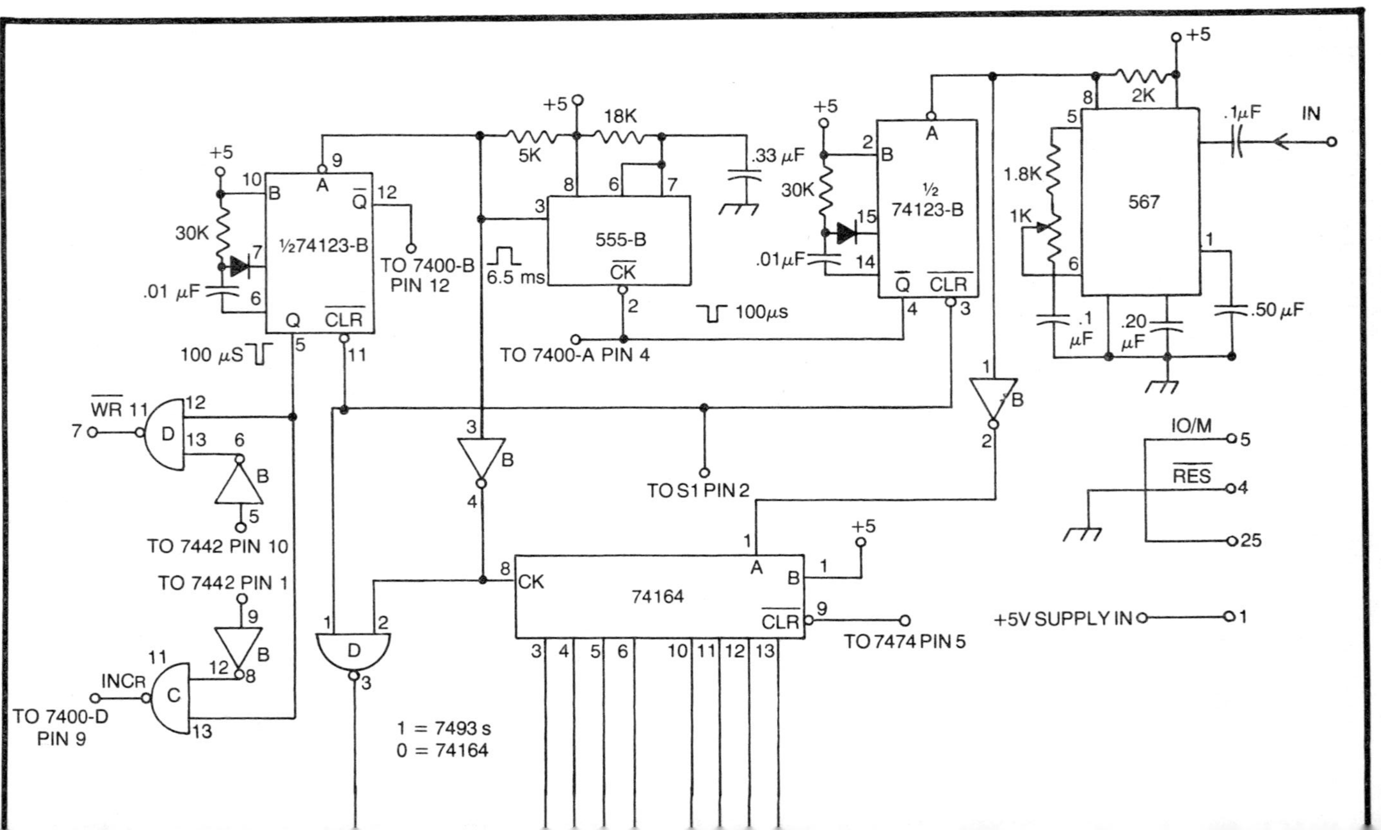

+5
2K
.1µF
IN
567
1.8K
1K
.50 µF
.1 µF
.20 µF
½ 74123-B
30K
.01µF
100µs
18K
5K
555-B
CK
6.5 ms
.33 µF
TO 7400-A PIN 4
½74123-B
TO 7400-B PIN 12
.01 µF
100 µS
WR 11
TO 7442 PIN 10
TO 7442 PIN 1
INCR
TO 7400-D PIN 9
TO S1 PIN 2
74164
TO 7474 PIN 5
1 = 7493 s
0 = 74164
IO/M
RES
+5V SUPPLY IN

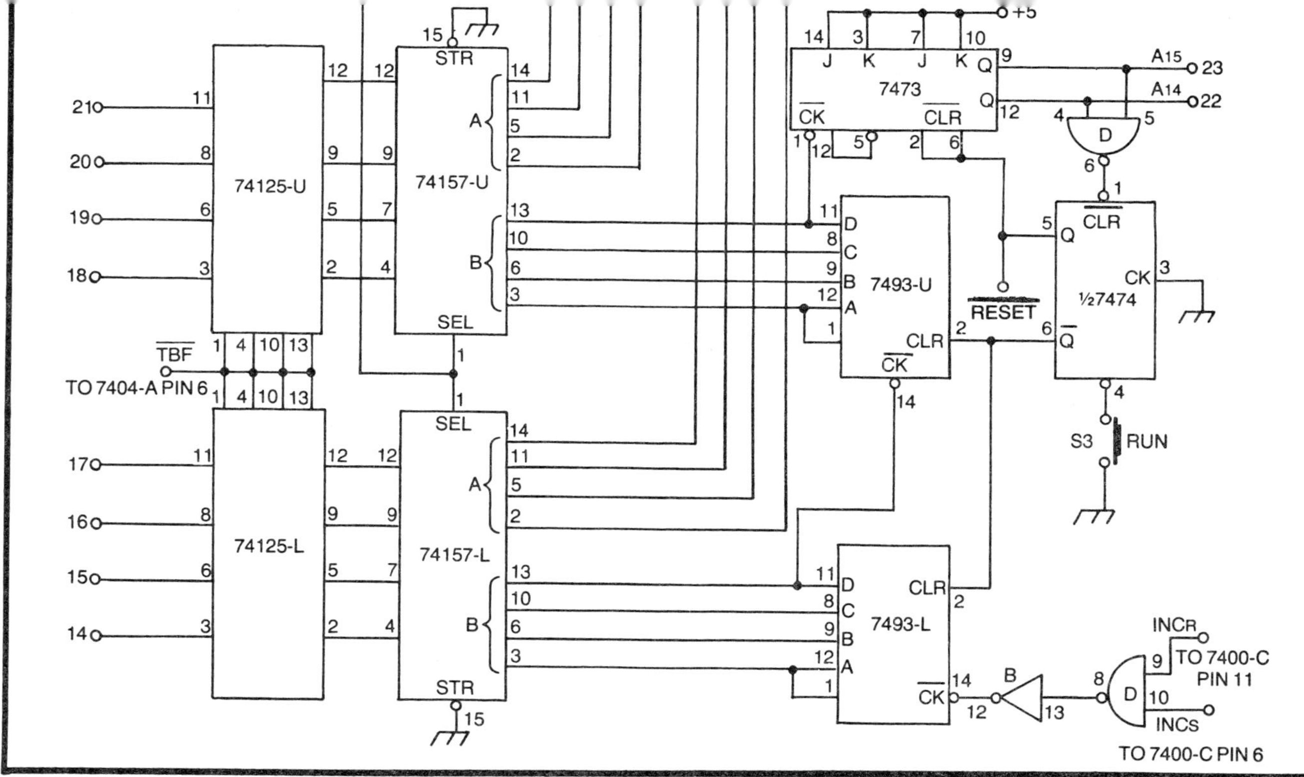

Fig. 14-7. Retrieval section of the tape interface. Diodes shown are 1N4148.

555-B output is also used, then, to control the SELECT pins of the data selectors which choose between address and data information for the AD lines. While the delay is running, the address counter is hooked to the AD lines, and the ALE line is pulsed. When the delay runs out, the shift register is hooked to the AD lines for writing into the RAM address.

How and when is the reassembled 8-bit word written into RAM? Here is where 7493-B and the 7442, shown back in Fig. 14-4, play their second role. Remember that 7493-B was used to keep track of the eight bits to be read in the storage mode. Now it will be used to keep track of the number of bits received, and thus, to indicate whether the data is ready for RAM storage. For this process, the second half of 74123-B is wired to produce a 100-microsecond pulse. This one-shot, however, is gated to trigger on the falling edge o the 555-B. Thus, when the 6.5-ms time delay runs out, this one-shot puts out a pulse.

This one-shot pulse does a few things. First, it is gated to clock the 7493-B of Fig. 14-4. Thus, each time a bit is received, and after the bit status has safely been latched into the shift register, the 7493-B advances one count to keep a record of the fact. The 7493-B begins normally at 0000, decimal zero. The first received bit clocks it to 0001, then upward until the eighth bit results in count #8, or 1000. At 1000, though, the 7442 decoder of Fig. 1-4 puts out a zero at its "8" pin. This is gated with the 74123-B pulse in Fig. 14-7 to produce a $\overline{WR}$ pulse. So as soon as the data is complete in the shift register, a $\overline{WR}$ pulse is issued, loading the word into the RAM location. (Since this $\overline{WR}$ pulse comes after the 555-B time delay, the 74157 data selectors are permitting the shift register data to reach the AD lines.)

Then will come that ninth pulse-tone on the tape, the one which is of irrelevant logical value—the one which the storage circuitry inserted purely for synchronization purposes. This pulse clocks 7493-B one more time, resulting in count #9, binary 1001; but the 7442 immediately resets the 7493-B to 0000 in this case. Now the 74123-B pulse is gated, using the 7442 pin labeled 0, so as to increment the address counter. In this way, the circuitry is ready for the next 8-bit word. The ninth tone, incidentally, *will* be analyzed and stored in the shift register, even though it is meaningless. But the subsequent 8-bit word shifts it out and it is harmlessly lost.

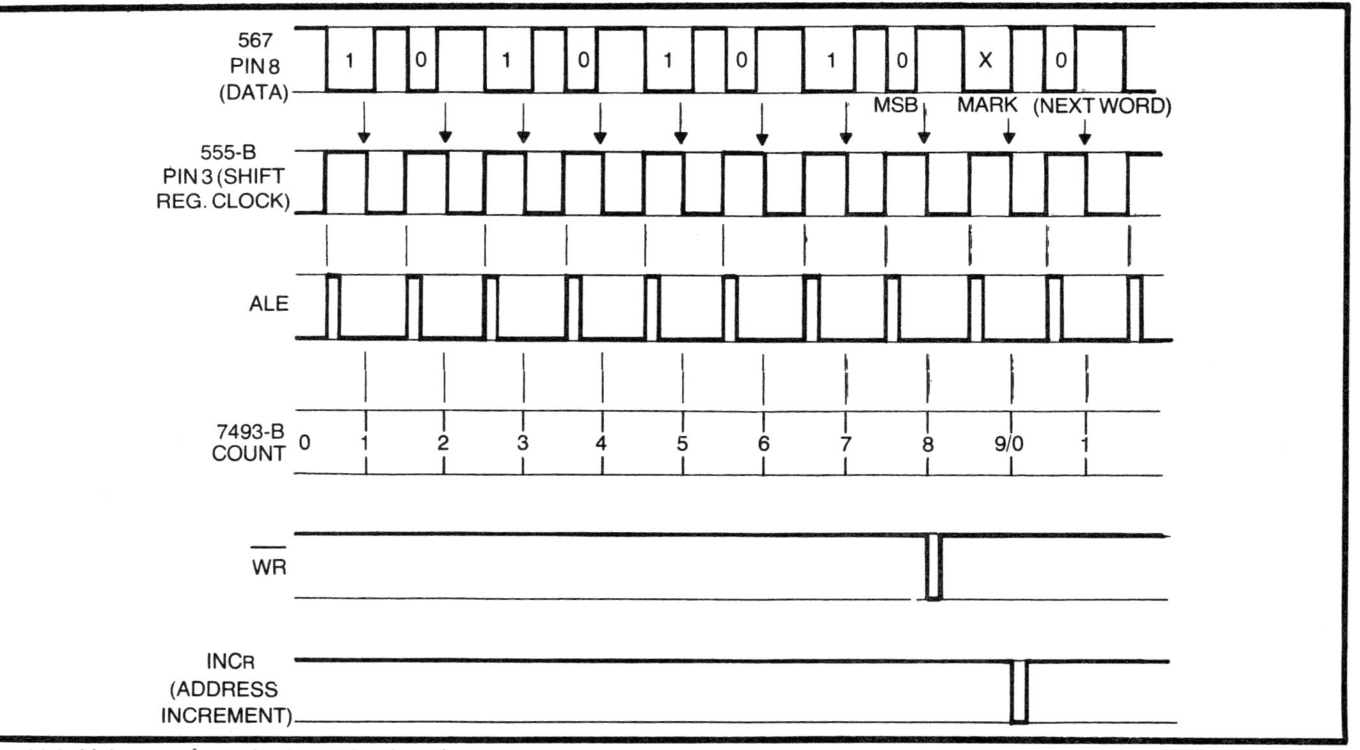

Fig. 14-8. Major waveforms for the retrieval section.

DISPLAY STAGE

We mentioned that, in the MANUAL mode of the storage process, the RAM locations can be manually accessed and examined using the STEP pushbutton. The display stage pictured in Fig. 14-9 comes into play here. Two 7447 *BCD-To-seven-segment decoders* are continually hooked up to the AD lines, and the results of their monitoring are displayed on a pair of segmented LED readouts. As was the case with the manual programmer of the preceding chapter, each 8-bit word is displayed in hexadecimal symbols; see Table 13-1 for the necessary conversion information.

When the interface is halted and in the storage mode, the LED readouts will be displaying the contents of RAM location zero. When the interface is running automatically, the displays become a blur until all memory has been stored; then location zero will once again be displayed. With switch S2 set to MANUAL, however, the process is controlled by the STEP pushbutton. As the interface is started by the RUN button, the display will not move from RAM address zero without pushing STEP. Each time STEP is pressed, the subsequent memory location is displayed. There are no provisions for address-reading—that must be kept track of by the user.

We saw in Fig. 14-5 that $\overline{RD}$ is only active less than half of the time. And so it is not enough merely to hook the display stage across the AD lines; it must be *strobed* so as to display only the output data from RAM. And so, in Fig. 14-4, the $\overline{RD}$ level is inverted and made available. This level is then used to strobe pins 4 of the 7447 display drivers. This results in a dimmer display than usual, but it is necessary.

CONSTRUCTION AND TESTING

As Fig. 14-2 demonstrated, the tape interface need not take up too much space if it is laid out carefully. The prototype interface board measures about 4.5″ by 5.0″, such economy being possible only through the medium of wire-wrapping techniques.

Once the entire unit has been constructed, it cannot be fully tested without some sort of program to store. But most of the individual circuits can be tested under artificial conditions. For instance, hook the interface up to a +5-volt supply, and plug the IN and OUT cords into a standard cassette recorder/player. The OUT cord should plug into the *auxiliary input* of the recorder, if there is

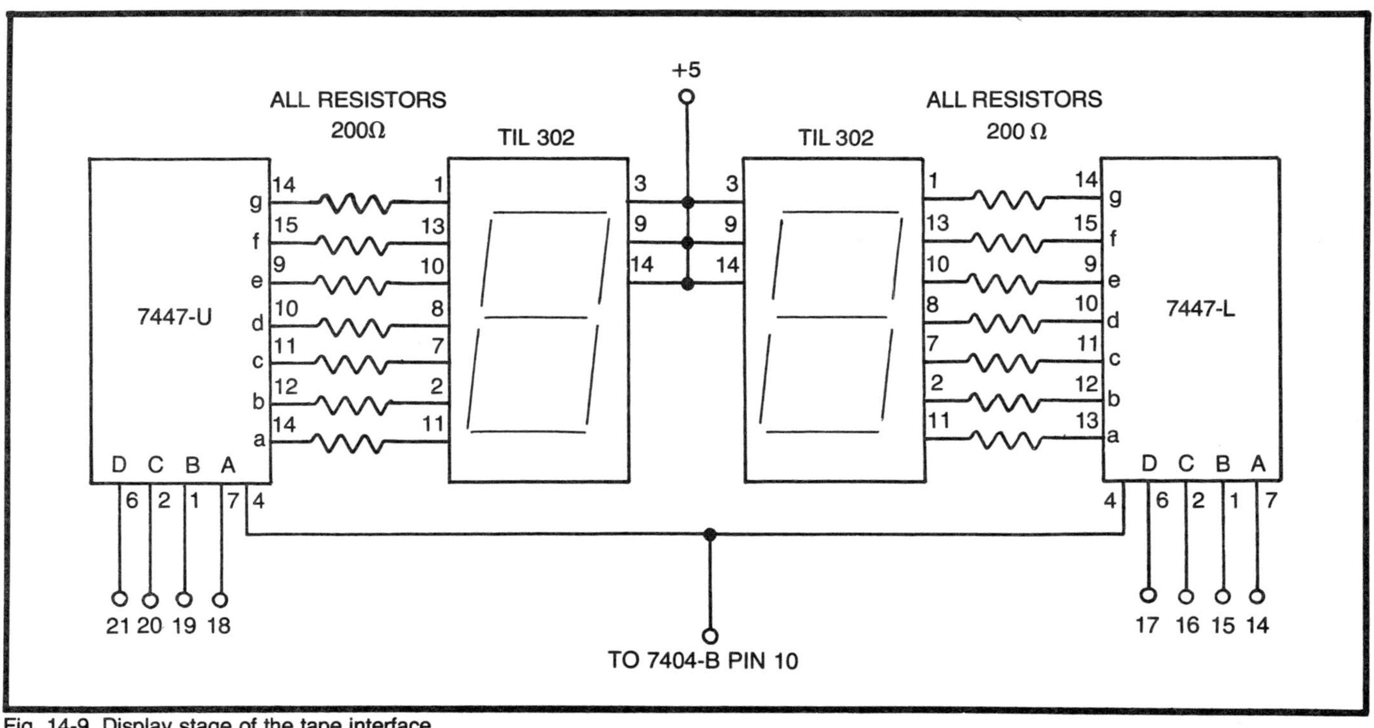

Fig. 14-9. Display stage of the tape interface.

one. Otherwise, the *microphone input* should be used, and the attenuation potentiometer shown at the output of Fig. 14-4 will have to come into play. Plug the IN cord into the *earphone jack* of the player, or its equivalent. Insert a good-quality cassette tape.

Set the switch S1 to STORE and S2 to AUTO. Then start the tape recorder in the *record mode*. As soon as you are certain that the cassette has passed any leader-tape, press RUN on the interface unit. Because the DB-25 is not plugged into anything, the store circuitry will read the input as an "all-ones" condition. That is, each 8-bit word it stores will be 11111111. This being so, do not expect the *display* to be active: the hexadecimal symbols for 11111111 would be two "blanks" (see Table 13-1).

Monitor the recorder if it has an input volume control; set this control for average recording levels, to avoid distortion which could adversely affect data storage. Monitor also, with an oscilloscope, the various waveforms which should be active in the storage circuitry. Trace the 40-kHz clock through the divider stages as far as the 100-Hz "baud rate" signal. Look for the ALE, $\overline{\text{RD}}$ and $\overline{\text{TBF}}$ signals according to Fig. 14-5. Check out the long and short pulses of 74123-A, and the clocking of 7493-B and the address counter. The entire process should last about 67 seconds, at which point the output tones will cease and all will reset.

Turn the cassette over and repeat the process, this time in the MANUAL mode, to test the debouncing stage. Monitor the output of the 558 when STEP is pressed; it should be a clean pulse. Monitor the address counter each time it is pressed, using either an oscilloscope or the *status display device* described in Chapter 10. The counter should not advance erratically or skip any numbers, or else the 558 is allowing some high-frequency "spikes" to re-clock the counter at unwanted times. Experiment some with the values of the network tied to pin 2 of the 558, or try a small capacitance from pin 1 to ground, if any problems are encountered. If the results are satisfactory, switch to AUTO and let the unit run its course until it stops.

Flip the cassette over again, and rewind it to the leader tape. Set the interface to RETRIEVE, and press RUN, then start the tape player in the *playback mode*. Monitor the output of the 567, which should be producing a pulse train, each pulse width being about 9 ms or less. Adjust the 1 K-ohm potentiometer at pin 6 of the 567 until the output pulses are as close to 9 ms as possible. Make note of the average pulse width of these pulses; it is the reference for a logical "1", and certain calibrations may be necessary.

Check the other levels of the retrieve circuitry. Watch the $\overline{\text{WR}}$ level, the ALE line, and the incrementation of 7493-B and the program counter. Confirm the delay of the 555-B, which should be on the order of 6.5 ms. This delay period may have to change, however, depending on the pulsewidth of the logical "1". If the "1" is close to or less than 6.5 ms, the circuit will not be able to differentiate between "1's" and "0's". Monitor one of the output pins of the 74164 and see if this is true: the pins should always be at a "1" level. If not, then the 567 pulses are finished by the time the 555-B delay times out. If so, *alter* the time delay at 555-B. Set it for approximately 75 percent of the average pulse width of a logical "1" as seen at 567 pin 8. If a "1" is only about 6 ms long, set the 555-B to about 4.5 ms, by altering the value of the resistor from pin 8 to pin 7.

The *display stage* cannot really be tested under normal conditions until the interface unit is in regular use. But at least it is possible simply to ground the various pins on the DB-25 connector, those correspondingly to the AD lines, while the unit is stopped and in the STORE mode. If you attempt this, however, unplug the two 74125 tristate buffer chips, so that the output of the address counter does not short to ground through the makeshift connections.

SUMMARY

By this point, the robot pet stands waiting to be programmed, to be given some sort of goal structure. To facilitate this process, two important programming tools have been developed. Now it is time to deal with programming itself, in depth and detail.

Before we can map out the software package for the robot pet though, we need to learn a bit about programming. The RCU-85, after all, does not operate in BASIC or FORTRAN, and certainly not in English. Before we can set the pet into independent motion, we must learn its language. The fundamentals of this language, and the principles which make it useful, are discussed one by one in the next chapter.

Chapter 15
Programming Techniques

The prime factor which makes this pet robot truly sophisticated is its basis in *software*. The pet could never be expected to approach as near to life-simulation as it does without its roots in a flexible, powerful and descriptive medium such as processor programming. A thorough grasp of the intricacies of the software control concept is, thus, a necessity if any success is expected in this project.

In the early days of computing, when the first "computers" came into being, there was very little flexibility. True, the machines of the late 1940's and early 1950's dealt on a binary numeric level, but all was for the most part "hardwired": data handling and storage was carried on by masses of mechanical relays. Vacuum tubes were next, and yet flexibility was low, and changes in processing were difficult. The advent of the transistor brought a revolution in size and power efficiency, but not yet in versatility.

It is impossible to separate the present microcomputer revolution, though, from the advent of mass-memory, particularly semiconductor memory. One could design and build some sort of the electronic controller which derived its pattern of procedure from a prewired circuit, but change would be impractical. However, if that pattern could be placed in a *symbolic* form into the bowels of some theoretical memory unit, it could be reorganized as quickly as the symbols could be rewritten! This very concept was merely awaiting the actual development of such a memory unit, and before long, *software* became synonymous with computing.

THE FEAR OF SOFTWARE

The layman's contact with software is usually not without much foreboding. He may walk into some formidable computing center and watch a programmer or technician typing obscure formulae into a huge business machine. The computer in turn spouts oracular answers beyond the ken of the bystander. Then there are things like punchcards, paper-tape, and long sheets of BASIC or FORTRAN listings, all calculated to defy translation.

In reality, if the truth be known, such things as programmer languages and punchcards are designed rather to *aid* the user in the day-to-day operation of the computer. They are attempts to convert the basically binary approach of the business machine into a form understandable to human beings—at least, those who are so trained!

The reader of this volume need not fear that he will have to enroll in his nearby community college for a semester of "Beginner's FORTRAN." The pet robot does not operate in such a high-level realm. The relatively simple operations of the pet do not demand such a perspective. Rather, the software for the pet is conceived and written in something more basic than BASIC: the *machine language*, the elementary binary code of the microprocessor.

The paradox, immediately obvious, is that is *should* be harder to approach software from this angle. After all, are not the high-level programming languages designed to place the processes of the computer into our *own* language? Why should we have to learn *its* language? The answer is related to a discussion we had earlier, in Chapter 7. The pet robot is basically *control-oriented*; it does not need very much programming or data to do what it needs to do. The imposition of a high-level language on it would be an inefficiency. (To put it in algebraic terms, it would take X units of memory to function as a pet, and 8X units merely to make it easier for the owner to write that first portion!) Moreover, high-level languages are oriented toward *thought* and conceptual situations. Low-level machine language is much more natural to use in the mundane world of closing relays and energizing servomotors.

In the world of the microprocessor, all such decisions of control are handled by binary words known as *instructions*. These instructions tell the processor essentially where to obtain relevant data, what to do with it, and where to place the results. Even as a controller, a processor does *not* think in terms of "Turn on relay four," but rather, "Move some data to a given position," where relay four is waiting. It is for this reason that all functions of the robot pet have been designed so that they can be initiated or terminated by the presence of either a +5 voltage or zero voltage. For the processor thinks only of data—binary ones and zeroes—but it handles them in the form of +5 or zero volts, a form to which the hardware developer can relate.

THE PROCESSOR AND INSTRUCTIONS

As we have seen, the robot pet is presided over by the RCU-85, a computer based on the Intel 8085A microprocessor. The 8085A has an instruction set of 80 possible instructions, each of which the processor reads from memory (if it has been written there previously by the human programmer) and executes. The instructions are actually made up of 8-bit binary numbers or "bytes." There are some instructions which are two bytes long (one after the other), and some three bytes long, but most are one byte in length. When these are organized into a *program*, a sequence designed to perform a goal, they are written sequentially in the memory of the RCU-85, which is capable of storing a long list of 8-bit bytes.

The 8085A instruction set may be divided into five essential groups, according to the kind of function the instruction brings about. These are:

- *Data Transfer Group*, for the moving and storage of data among the internal registers and memory;
- *Arithmetic Group*, for the performing of arithmetic functions (addition, subtraction, incrementation) on data in the internal registers and in memory;
- *Logical Group*, for the performing of logical operations (and, or, comparison, and so on) on data in the internal registers and in memory;
- *Branch Group*, for the interruption of the normally-sequential program order (conditional jumps through memory);
- *Stack, I/O, Machine Control Group*, for operations involving I/O ports, the 8085A interrupt system and address-storage stack in memory, and various processor functions.

Figure 15-1 shows a conceptual diagram of the interactive elements of the RCU-85 system. The 8085A processor itself has a number of inherent features, which must be described before programming can be dealt with effectively. (Note, though, that the breakdown given in Fig. 15-1 is for explanatory purposes relating to the robot pet system only. Many internal features of the chips have been omitted for simplicity).

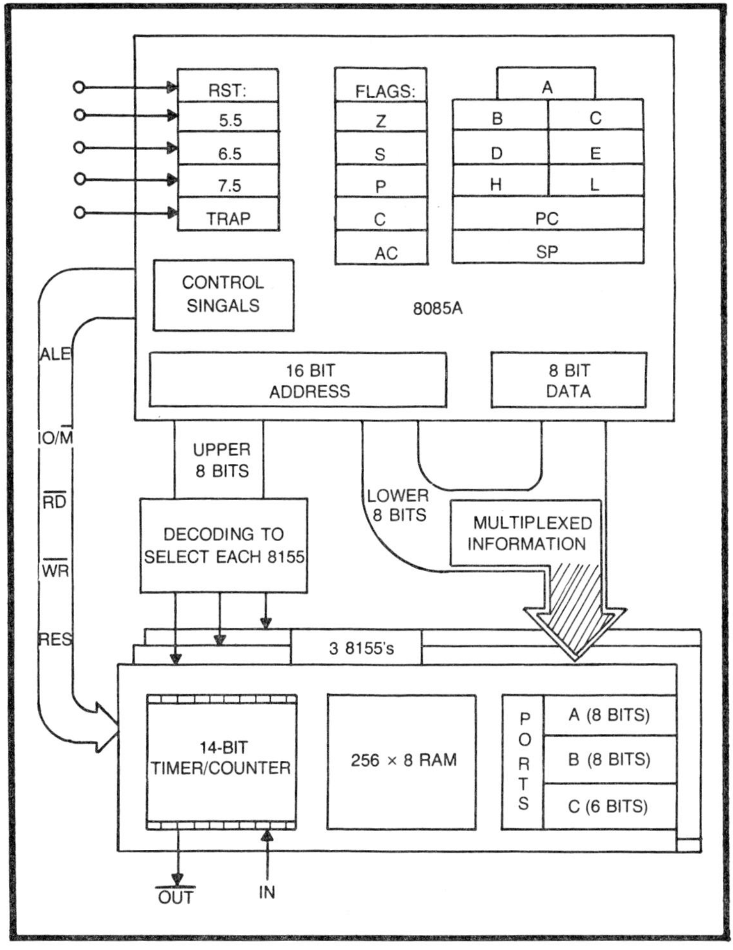

Fig. 15-1. Interactive elements of the RCU-85 from a programming standpoint.

First, the 8085A has a set of seven "general purpose" registers, labeled A, B, C, D, E, H, and L. These are merely 8-bit internal temporary-storage locations, to facilitate the execution of certain instructions. One of the seven, A, is of primary importance; it is also sometimes called the *accumulator*, and it is used more than the others in most of the instructions.

Notice, too, that some of these registers are pictured as paired. This is to indicate that, in certain instructions, the processor can use these paired registers as if they were double-sized, 16-bit registers. (This makes it easier to temporarily store the addresses of memory locations, which are 16-bits in length.) B-C, D-E, and

H-L are these register pairs, with H-L being the pair most frequently used in the instruction set.

In addition to the 8-bit registers, there are two internal 16-bit registers, one termed the *program counter* (PC), and the other the *stack pointer* (SP). The program counter is mainly a 16-bit binary counter which, starting at "all zeroes," counts upward, issuing a 16-bit address out to the RCU-85 memory. In this fashion, the 8085A reads each sequential location in memory, executing each instruction it happens to find. Of course, if the instruction tells the 8085A to *jump* to some far off memory location, the 8085A will *load* the program counter with that new address, and proceed to read sequentially from the new address.

The stack pointer is a part of the interrupt system of the 8085A, which was briefly described in Chapter 7. The *interrupt system* is designed so that some external device can temporarily side-track the 8085A into some other task, and then return it to its original program without any perceptible change or upset. (As we saw in Chapter 7, there are four special interrupt pins which trigger this process.) The purpose of the stack pointer is to help preserve the original status of the program flow of the 8085A *before* the interruption, so that a return to normalcy can be made without confusion.

Here's how it's done. The programmer, before he writes his programs, sets aside a certain portion of memory, which is termed the *stack,* just for the purpose of storage of addresses or data which need temporary preservation during an interrupt sequence. Then, at the beginning of his program, he uses a certain instruction to load the stack pointer with the *last address* of that portion of memory, so the 8085A "knows" what it is free to use. Then whenever the processor is interrupted in the middle of some program, it *automatically* stores the next address of the regular program in this previously-designated portion of memory, so that it can later pick up where it left off. This is a simplified understanding of the process, but it will suffice until we discuss the setting up of the robot pet's stack later on.

In addition to the internal registers we've just described, there are also a set of internal single-bit flip-flops, or *flags*, built in to help keep track of the result of certain arithmetic or logical operations. Any one of these flags can be tested by an instruction, so that the 8085A can choose a course of action based on the result of, say, an addition or comparison. These all-important flags are:

- *Zero Flag* (Z), which is set to a 1 if the result of a previous instruction resulted in an answer of 00000000;

- *Sign Flag* (S), which is set to a 1 if the most significant bit of the result of an instruction is a 1 (i.e., if the result is 1xxxxxxx, where x is irrelevant);
- *Parity Flag* (P), which is set to a 1 if the sum of the "1's" in the result of an instruction is even;
- *Carry Flag* (C), which is set to a 1 if the result of an instruction brought about a carry or a borrow (arithmetically) out of the highest bit;
- *Auxiliary Carry Flag* (AC), which is set to a 1 if the result of an instruction brought about a carry out of bit 3 and into bit 4.

As far as the robot pet programming is concerned, only a few of these flags need concern us. The Zero and Carry flags are both important, since they are used in instructions which help to identify incoming data, such as a Comparison instruction. The Sign flag and the Auxiliary Carry flag are used for medium-level arithmetic, which is not necessary in the robot pet. The Parity flag is used in error detection for serial data transmission, which will not be handled in the software of the robot pet. The use of the flags will be further illumined by the *instruction set*.

THE INSTRUCTION SET

The following body of information is the *8085A instruction set* in its entirety, for reference purposes. (This material is provided courtesy of Intel Corporation, based on their *MCS-85 User's Manual*, copyright 1978.) Each instruction has a mnemonic, an 8-bit binary machine code, and an explanation. If an instruction has a certain effect on the flags, the explanation will note this; otherwise, the general rule for effect on flags is given for each major instruction grouping.

In some of the machine codes, a group of three bits may be unspecified and labeled with the variable SSS or DDD. Such a label designates one of the internal registers which is to be affected by the instruction, according to the following code:

Table 15-1. DDD And SSS Codes

DDD or SSS	REGISTER
111	A
000	B
001	C
010	D
011	E
100	H
101	L

SSS indicates that the register chosen is to be the SOURCE of data in an operation, and DDD represents the register to be the DESTINATION of data. Thus, in the MOV$_{r1,r2}$ instruction, coded 01DDDSSS, data is moved from Register SSS to Register DDD, whatever these are designated to be.

There are also a few cases in which the letters RP take the place of two bits in a machine code. These are used to represent any register-pair or the stack pointer, according to the code in Table 15-2.

Table 15-2. RP Codes

RP	REGISTER-PAIR
00	B-C (B is upper, C is lower)
01	D-E (D is upper, E is lower)
10	H-L (H is upper, L is lower)
11	SP

A final case in which a variable field is used is in the ease of the conditional instructions of the *branch group*. In these, a three-bit field termed CCC indicates the condition under which the instruction will take place. Basically, the conditions are related to the states of the various flags, Z, C, P, and S.

Table 15-3. Condition & Codes

CCC	CONDITION
000	NZ-not zero (Z = 0)
001	Z-zero (Z = 1)
010	NC-not carry (C = 0)
011	C-carry (C = 1)
100	PO-parity odd (P = 0)
101	PE-parity even (P = 1)
110	P-plus (S = 0)
111	M-minus (S = 1)

Most of the instructions are self-explanatory, but a few points should be noted beforehand. First, when data is *moved* into some location, it is being *duplicated*; that is, the data does not cease to exist in the original location. And so, to say that "data is moved from A to B" really means that data in A will now be present in *both* A and B. Second, the terms *and, or, inclusive* and *exclusive* are used in the *logical group* without definition; the reader should consult a dictionary or logic text for the standard definitions.

Table 15-4. Data Transfer Group—No Flags Are Affected

Instruction	Code	Description
$MOV_{r1,r2}$	01DDDSSS	Moves data from register SSS to register DDD
$MOV_{r,M}$	01DDD110	Moves data from the memory location whose address is in H-L to register DDD
$MOV_{M,r}$	01110SSS	Moves data from register SSS to the memory location whose address is in H-L
$MVI_{r,data}$	00DDD110 DATA BYTE	Moves 2nd byte into register DDD
$MVI_{M,data}$	00110110 DATA BYTE	Moves 2nd byte into the memory location whose address is in H-L
$LXI_{rp,data}$	00RP0001 LOWER DATA UPPER DATA	Moves 2nd byte into lower register of register-pair RP; moves 3rd byte into upper register of register-pair RP
LDA_{addr}	00111010 LOWER ADDR UPPER ADDR	Moves data from memory location addressed by 2nd and 3rd bytes to register A
STA_{addr}	00110010	Moves data from register A to memory location addressed by 2nd and 3rd bytes
$LHLD_{addr}$	00101010 LOWER ADDR UPPER ADDR	Moves data from memory location addressed by 2nd and 3rd bytes to register L; moves data from the next location to register H
$SHLD_{addr}$	00100010 LOWER ADDR UPPER ADDR	Moves data from register L to memory location addressed by 2nd and 3rd bytes; moves data from register H to next memory location
$LDAX_{rp}$	00RP1010	Moves data from the memory location whose address is in register-pair RP to register A (Pair H-L may not be selected for RP)
$STAX_{rp}$	00RP0010	Moves data from register A to the memory location whose address is in register-pair RP (Pair H-L may not be selected for RP)
XCHG	11101011	Exchanges data between registers H and D; exchanges data between registers L and E

USING THE INSTRUCTION SET

The sheer size of the description of the instruction set may frighten you, but remember that the largest, most complex programs are simply combinations of these 80 instructions. It is possible to write a program only six or seven steps along which would replace the control capabilities of several solid-state parts. The ultimate version of the robot pet's software is just a collection of dozens of such simple mini-programs. Just to prove the point, let's look at a number of such techniques, to show how we can make the instruction set work for us.

Table 15-5. Arithmetic Group—Unless Indicated, All Flags Are Affected

ADDr	10000SSS	Adds register A to register SSS; results are placed in register A
ADDM	10000110	Adds register A to content of memory location addressed by H-L; results placed in register A
ADIdata	11000110 DATA BYTE	Adds register A to 2nd byte; results are placed in register A
ADCr	10001SSS	Adds register A, register SSS, and Carry; results are placed in register A
ADCM	10001110	Adds register A, contents of memory location addressed by H-L, and Carry; results are placed in register A
ACIdata	11001110 DATA BYTE	Adds register A, 2nd byte, and Carry; results are placed in register A
SUBr	10010SSS	Subtracts register SSS from register A; results are placed in register A
SUBM	10010110	Subtracts contents of memory location addressed by H-L from register A; results are placed in register A
SUIdata	11010110 DATA BYTE	Subtracts 2nd byte from Register A; results are placed in register A
SBBr	10011SSS	Subtracts register SSS and Carry from register A; results are placed in A
SBBM	10001110	Subtracts contents of memory location addressed by H-L and Carry from register A; results are placed in register A
SBIdata	11011110 DATA BYTE	Subtracts 2nd byte and Carry from register A; results are placed in register A
INRr	00DDD100	Register DDD is incremented; Carry not affected
INRM	00110100	Contents of memory location addressed by H-L are incremented; Carry not affected
DCRr	00DDD101	Register DDD is decremented; Carry not affected
DCRM	00110101	Contents of memory location addressed by H-L are decremented; Carry not affected
INXrp	00RP0011	Register-pair RP is incremented; no flags are affected
DCXrp	00RP1011	Register-pair RP is decremented; no flags are affected
DADrp	00RP1001	Register-pair RP is added to H-L; results are placed in H-L; only Carry flag is affected
DAA	00100111	Contents of register A are adjusted to form two 4-bit BCD digits

Table 15-6. Logical Group—Unless Indicated, All Flags Are Affected.

ANA_r	10100SSS	Register SSS and register A are "and-ed"; results are placed in A; Carry is cleared, AC flag is set
ANA_M	10100110	Contents of memory location addressed by H-L and register A are "and-ed"; results are placed in A; Carry is cleared, AC flag is set
ANI_{data}	11100110 DATA BYTE	2nd byte and register A are "and-ed"; results are placed in A; Carry is cleared, AC flag is set
XRA_r	10101SSS	Register SSS and register A are "exclusively-or-ed"; results are placed in A; Carry and AC flag are cleared
XRA_M	10101110	Contents of memory location addressed by H-L and register A are "exclusively-or-ed"; results are placed in A; Carry and AC flag are cleared
XRI_{data}	11101110 DATA BYTE	2nd byte and register A are "exclusively-or-ed"; results are placed in A; Carry and AC flag are cleared
ORA_r	10110SSS	Register SSS and register A are "inclusively-or-ed"; results are placed in A; Carry and AC flag are cleared
ORA_M	10110110	Contents of memory location addressed by H-L and register A are "inclusively-or-ed"; results are placed in A; Carry and AC flag are cleared
ORI_{data}	11110110 DATA BYTE	2nd byte and register A are "inclusively-or-ed"; results are placed in A; Carry and AC flag are cleared
CMP_r	10111SSS	If data in A = data in SSS, Z flag is set; If data in A > data in SSS, Carry is set;
CMP_M	10111110	If data in A = data in memory location addressed by H-L, Z flag is set; If data in A > data in memory location addressed by H-L, Carry is set;
CPI_{data}	11111110 DATA BYTE	If data in A = 2nd byte, Z flag is set; If data in A > 2nd byte, Carry is set;
RLC	00000111	Register A is shifted left one place; highest bit is shifted around to lowest position and into Carry
RRC	00001111	Register A is shifted right one place; lowest bit is shifted around to highest position and into Carry
RAL	00010111	Register A is shifted left one place; highest bit is shifted into Carry, and Carry is shifted into lowest bit
RAR	00011111	Register A is shifted right one place; lowest bit is shifted into Carry, and Carry is shifted into highest bit
CMA	00101111	Register A is complemented (all bits reverse state); no flags are affected
CMC	00111111	Carry is complemented (reverse state); no other flags are affected
STC	00110111	Carry is set to 1; no other flags are affected

Table 15-7. Branch Group—No Flags Are Affected

JMP_{addr}	11000011 LOWER ADDR UPPER ADDR	Program jumps to address specified in 2nd and 3rd bytes
$Jcond_{addr}$	11CCC010 LOWER ADDR UPPER ADDR	The conditions of JMP_{addr} occur if condition CCC holds true
$CALL_{addr}$	11001101 LOWER ADDR UPPER ADDR	Program jumps to address specified in 2nd and 3rd bytes; the next address after CALL is stored in the two memory locations addressed by SP-minus-1 and SP-minus-2; SP is decremented twice
$Ccond_{addr}$	11CCC100 LOWER ADDR UPPER ADDR	The conditions of $CALL_{addr}$ occur if condition CCC holds true
RET	11001001 LOWER ADDR UPPER ADDR	Program jumps to the address stored in the two memory locations addressed by SP and SP-plus-1; SP is incremented twice
Rcond	11CCC000	The conditions of RET occur if condition CCC holds true
RST_n	11NNN111	Program jumps to the address NNN-times-8; the next address after RST is stored in the two memory locations addressed by SP-minus-1 and SP-minus-2; SP
PCHL	11101001	Program jumps to the address stored in H-L

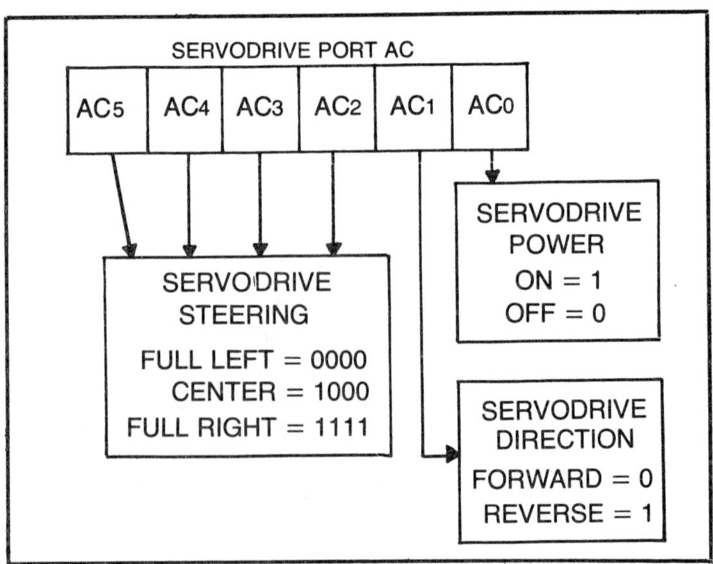

Fig. 15-2. Servodrive port layout, for the purposes of illustration.

Table 15-8. Stack, I/O, Machine Control Group—Unless Indicated, Flags Are Not Affected

PUSHrp	11RP0101	Contents of register-pair RP are moved to the memory locations addressed by SP-minus-1 and SP-minus-2; SP is decremented twice; (SP may not be selected for RP)
PUSHPSW	11110101	Register A and the flags are stored in the memory locations addressed by SP-minus-1 and SP-minus-2; SP is decremented twice;
POPrp	11RP0001	Contents of the two memory locations addressed by SP and SP-plus-1 are moved to register-pair RP; SP is incremented twice; (SP may not be selected for RP)
POPPSW	11110001	Contents of the two memory locations addressed by SP and SP-plus-1 are moved to register A and the flags; SP is incremented twice
XTHL	11100011	Contents of the two memory locations addressed by SP and SP-plus-1 are exchanged with the contents of H-L
SPHL	11111001	Contents of H-L are moved into SP
INport	11011011 PORT ADDR	Data is moved from the port addressed by the 2nd byte into register A
OUTport	11010011 PORT ADDR	Data is moved from register A into the port addressed by the 2nd byte
EI	11111011	Interrupt system is enabled after the next instruction is executed
DI	11110011	Interrupt system is disabled immediately
HLT	01110110	Processor is halted; registers and flags are unaffected
NOP	00000000	No operation is performed; registers and flags are unaffected
RIM	00100000	Status of interrupts, masks, and SID pin are moved to register A
SIM	00110000	Interrupt masks and SOD pin are set up from register A

Let's say, for instance, that we want to make the pet robot move. As we already know, the lines that control the motorframe relays and the servodrive circuitry are linked to 8155-A, port AC, as shown in Fig. 15-2. Now, if we want to make the pet robot move straight forward, we want three things done:

- Bit 0 must be a 1, to energize the drive motor.
- Bit 1 must be a 0, to permit forward direction.
- Bits 2-5 must be 1000, to steer directly forward.

In short, we need to place the number 00100001 at port AC of 8155-A. (The two zeros of bits 6 and 7 will be ignored by the port chip, because port AC is only a 6-bit port.)

To do this, we need to write the number 00100001 into register A, then output it from A to the proper ports. This is a simple process. All it takes is a *Move Immediate* (MVI) instruction to place the data into register A. The DDD code of the Move Immediate instruction is replaced by the code 111, which indicates that the data is to be placed into A. The second byte of the Move Immediate instruction is the data to be stored. So the first step would look like this:

ADDRESS	INSTRUCTION	CODE	Explanation
0	MVI A, data	00111110	Data moved to A
1		00100001	

Now that the data is in A, it's easy to get it out of the port with the *output* (OUT) instruction. The port address for 8155-A port AC is 00000011, which becomes the second byte of the output instruction. This is added to the earlier step:

0	MVI A, data	00111110	
1		00100001	Data moved to A
2	OUT port	11010011	
3		00000011	Data from A to port

Thus, with only two instructions and four bytes, we've commanded the robot pet. Because the 8155 chips have latched outputs, the outputted data will remain valid until instructed to change or until the RCU-85 is reset. That is, the robot pet will keep on rolling forward, just as it has been told.

By changing the data which was loaded into A, the program is changed. Thus, if the number 00100011 were in address 1 as the second byte of the MVI instruction, the robot pet would move straight in reverse.

This is impressive, except that it's hardly a good analogy of living things. People, when they walk, do not close their eyes and take off, full speed forward, oblivious to surroundings. They condition their motion by intelligent decision-making. Likewise, pet robot must scan the surroundings with its Soniscan abilities and be careful to avoid obstacles.

The discussion on the full workings of the Soniscan software yet awaits us, and so we'll not use it as our next example. But suppose, instead, that the robot dog had a photoelectric cell mounted in its forehead; that when a bright light hit the cell, its circuit would generate +5 volts (logical 1); and that the output of this circuit were tied to 8155-A, port AB, bit 7. Furthermore, let's say that the designer wished the pet robot to roll forward, as in the previous example, *until* it saw a bright light. At such a time, it should stop dead in its tracks.

The first four address locations of the program would be the same as already described: a Move Immediate and an Output to get the pet rolling. Next, though, the RCU-85 needs to interrogate the proper port, to see if a light has been sighted yet. For this, an Input instruction is used. The port address for 8155-A port AB is 00000010; and so the instruction would appear like this:

4	IN port	11011011	Data from port to A
5		00000010	

This instruction reads the entire 8 bits of the port and places that data into register A.

Now we have a small hurdle to jump. We *could* use some sort of comparison instruction to find out if that all-important bit 7 is a 1 or a 0; but what about the rest of the bits? If anything else is hooked to those bits at the port, then those bits could be 1's or 0's, and unless we know, we can't use just a Comparison instruction. We need some way to isolate bit 7 and make all the other bits in A into zeroes.

The way to do this is with the AND Immediate Instruction (ANI). This useful statement logically ANDs register A with its own second byte, placing the results in A. And by the very nature of the logical AND operation, if a zero is ANDed with *any* number, it will result in a zero. It is possible, then, to write the second byte of such an instruction as 10000000; if this number is ANDed with register A, all bits except for bit 7 will be zeroed. Thus, bit 7 is isolated for testing purposes, unaffected by the ANI instruction.

6	ANI data	11100110	AND data with A
7		10000000	

Now we can use a Compare Immediate instruction. The CPI instruction compares A to its own second byte, and sets the Z flag to a 1 if the two words are the same. This flag can later be tested, and a decision can be made whether or not to stop. We want to know if bit 7 is a 1 or not, so register A should be compared to the number 10000000. If the light hits the photoelectric cell, bit 7 will go to a 1, and the successful CPI instruction will flip the Z flag to a 1.

8	CPI data	11111110	Compare data to A
9		10000000	

Finally we can test the state of the Z flag to see if the light has been sighted. We want a loop to be set up; that is, we want a jump instruction that will loop the program back to address 0 if no light is sighted. That way, as long as no light is sighted, the robot pet will continue to repeat addresses 0 through 9 and never change; it will always move forward. If a light *were* seen, though, the jump would not occur, and the program would go to the next location. (And presumably, in that next location would be an instruction to tell the pet to stop.)

So we need an instruction which will cause a jump if *no* light is seen; that is, if the Z flag has *not* become a 1. Of the CCC conditions listed at the beginning of the instruction set, this is condition NZ, or Not Zero, with a code of 000. (Note that "Not Zero" means *"Zero* flag has *not* flipped to a one." Be careful not to confuse the term "Zero" with the state of the port pin.) The instruction, then, is *Jump if Not Zero*, (JNZ) with the second and third bytes making up a 16-bit address of the location to which to be jumped. (We wanted to return to address 0, which would be represented by sixteen zeroes.)

10	JNZ	11000010	Jump if Not Zero
11		00000000	to address 0
12		00000000	

The final step is to place a statement after the JNZ instruction, which tells the robot pet to stop. Ordinarily, when no light is seen, the program would never reach the statement, caught hopelessly in the loop. But the light would nullify the JNZ statement, allowing the *stop* command to be read and executed. The stop command would be identical to the *start* command of addresses 0 through 3, except that the second byte of the MVI instruction is changed to reflect the new goal. The entire program, then, would look like this:

0	MVI A, data	00111110	Data moved to A
1		00100001	
2	OUT port	11010011	Data from A to port
3		00000011	
4	IN port	11011011	Data from port to A
5		00000010	
6	ANI data	11100110	AND data with A
7		10000000	
8	CPI data	11111110	Compare data to A
9		10000000	
10	JNZ addr	11000010	Jump if Not Zero
11		00000000	to address 0
12		00000000	
13	MVI A, data	00111110	Data moved to A
14		00100000	
15	OUT port	11010011	Data from A to port
16		00000011	

Reread the explanation behind the previous program and get a good grasp of the concept behind it. In essence, the pet robot software package is not much more than an interwoven tapestry of many such *subroutines*. The instructions above are among the most frequently used in the programming to come.

FURTHER INSTRUCTIONS

A number of other useful instructions are in the instruction set, a few of which it may be worthwhile to point out. There are, for instance, the four *Rotate* instructions, RLC, RRC, RAL, and RAR. Essentially, these instructions use register A as if it were a shift register. All of the bits are shifted left or right one bit-position. Depending on the instruction, the bit on the end is either shifted around to the other end, or is shifted into the Carry flag as a sort of "ninth bit." These instructions are useful if the programmer wants to organize a number of bits into a specific position. Let's say, for example, that port BB is an output port, and that the lowest bit, BB_0, is hooked to a beeper. A sequence of beeps and spaces can be generated simply by loading A with a certain eight-bit word, then repeatedly rotating the word right (RAR) and outputting the word each time (OUT). Naturally, for the pattern to be audible, some time delay must be interpolated, since the instructions are executed so rapidly. But the concept is clear enough.

Another extremely useful instruction is CALL, in the branch group. Suppose there is a special routine which is used very often throughout the larger program but is so long that it's impractical to repeat it each time. It would waste too much memory space. The answer is to write the routine once in memory, starting at a certain address. Then each time it is needed, use a CALL instruction, specifying the beginning address of the routine in the second and third bytes of the CALL. The program jumps to the routine, keeping track of where it came from using the *interrupt stack*. At the end of the routine is a *return* instruction (RET), which causes the program to check the interrupt stack to find the "way home".

The 8085A instruction set is one of the most powerful among microprocessors offered today. There are instructions use the registers as counters to keep track of events, such as INR. There are instructions that modify the very contents of the RAM memory itself, such as STA. The program can jump to any memory location using JMP. The program can even "freeze" itself using HLT, refusing to continue until the processor is reset. The wide variety of instructions, coupled with the simplicity of the tasks involved, make it possible to write a tightly-packed, "memory-stingy" software package for the robot pet.

As always, the clearest manner in which to explain an idea is in application. And so the proof of the power of the 8085A instruction set lies in MECH 1, the software for the pet. The interrelation of every function of the pet robot, and the techniques behind each step, are treated in the next chapter.

Chapter 16
Mech 1: The Software Package

You have traveled a long, hard road to reach this chapter. You've formed metal, wrapped wire and tested circuits. And most recently, you have exercised your mind in the logical discipline of simple machine-code programming techniques. The final step now awaits you, and that is the laborious and careful construction of the software package for the robot dog.

It is important to be reminded that the secret to the programming of the pet is the clever use of the powerful 8085A instruction set. All great programs consist of many, many smaller ones, and the heart of these are the individual instructions. You should begin your task with a good healthy dose of review of the previous chapter, and perhaps even a bit of trial-and-error programming of your own on paper. The more familiar the instructions become, the more natural it will be to use them effectively.

Another important tool in the task of programming is the practice of neat, comprehensive documentation. As the ideal full-size program begins to unfold, certain portions—and the rationale behind them—will grow obscure in the mind. A little time spent in noting the purpose of each instruction or subroutine will be of inestimable value in debugging of the program package as a whole.

To facilitate this sort of discipline, a "programming form" such as the one illustrated in Fig. 16-1 is helpful. At the top of the sheet, there is room for broad information, including the page number and the individual RAM chip (of the three) being addressed. Columns are then provided for the address (in decimal), the hexadecimal address, the instruction mnemonic, the eight-bit instruction code, the hexadecimal equivalent of the instruction code, and explanatory information concerning the program. In this manner, all relevant data is kept orderly and accessible. The hexadecimal columns, incidentally, need not be filled in until actual programming of the RCU-85 is to occur; the *manual programmer*, remember, handles all addresses and instructions as hexadecimal figures on its display. Note, too, that the symbols for numbers 10-15 in the hexadecimal columns are those which appear on the seven-segment readout, rather than the customary A, B, C, D, E, F notation. This is a matter of taste rather than necessity, although it has the advantage of eliminating continual "translation" when loading the program.

The software package described in this chapter is termed MECH 1, and is a good first standard program for the robot canine. You may choose to write an entirely different sort of package; nevertheless, the same basic elements will need to be included. The detailed description that follows will give some idea of the number of factors involved. As each step is discussed, refer to the software listings located at the end of this chapter. It is impractical to explain each and every instruction, but what is not detailed here can surely be derived from a careful review of the 8085A instruction set.

INITIALIZATION

When the pet is first activated, brought out of reset, the RCU-85 immediately begins at *address zero* in memory, which is located in 8155-A. Now, when the RCU-85 starts up, the port sections of the 8155s are configured as input ports. The internal timers, also, are "unconfigured." So to set up the ports and timers according to our prescribed plan (see Fig. 7-5), the beginning of the program must contain an *initialization* routine, which programs the 8155s. Once these ports and timers are configured, they will remain in the desired modes until a reset occurs, or (if desired for some reason) until reconfigured by a new initialization routine. There are a few other functions of the initialization routine, which will be described, but this matter of the ports and timers is primary.

The initialization routine runs from address 0000 to 001D in the RAM of 8155-A. The first object of interest is the configuration of the 8155-B timer. This is the counter which, as you will recall from Chapters 7 and 9, operates as a "divide-by-256" counter and a time reference for the Soniscan subsystem. Figure 7-6 gives the set-up of the two registers in the 8155-B, TMR HI and TMR LO, which relate to the counter. In order to make the counter into a "divide-by-256" device, the number 256 must be loaded into these regis-

PAGE ___ OF ___ PAGES

DEC ADDR	HEX ADDR	MNE	OP CODE	HEX CODE	COMMENTS
0	00				
1	01	MOV	01111000	78	DATA B→A
2	02	ADI	11000110	⊔6	DATA + A
3	03		01100100	64	(DATA=64HEX)
4	04	OUT	11010011	⊑3	A→PORT
5	05		01000011	43	PORT#34HEX
6	06	IN	11011011	⊑⊐	PORT→A
7	07		00000001	01	(PORT #01HEX)
8	08	ANI	11100110	ⱡ6	DATA ∧ A
9	09		10000000	80	(DATA=80HEX)
10	0⊏	MOV	01000111	47	DATA A→B

Fig. 16-1. A programming form for organization of software documentation.

ters, along with the mode code for a repeating output square wave, which code is 01. Thus, as Chapter 9 described, the number 01000001 should be loaded into TMR HI, and 00000000 into TMR LO.

To do this, an MVI instruction is used to load each 8-bit word into 8085A register A; then an OUT instruction takes the data from A and sends it into the 8155 chip. The addresses 44 and 45 (in hexadecimal) are used in the second byte of the OUT instructions to indicate the timer registers. RAM addresses 0000 through 0007 (hex) contain these steps.

Next come the ports of the 8155s. Figure 7-7 shows the command/status (C/S) register of the 8155, which is loaded to set up the ports and to start the counters. The initialization routine begins by loading C/SA, which has an OUT address of 00 (hex). As Fig. 7-5 indicates, we desire port AA to be input, port AB to be input, and port AC to be output. Thus, according to Fig. 7-7, the number to load is 00001100. The upper two bits would be used to start the timer, but the timer in 8155-A is only used on occasion by the program later on...so there is not need to start it now.

C/SB is the same as C/SA in port configuration, if we choose arbitrarily to set port BB, unused, to the input mode. But in this case, we *do* want to start the timer, which has already been set up in steps 0000 to 0007; it is to be running constantly from the very beginning. And so, the upper two bits of the C/SB register must be set to 11, which starts the counter running. The mode of the 8155-B timer, remember is a continuous square-wave output, and so it will keep running from now on, until told to stop, or until a reset occurs.

C/SC need not be loaded, of course, if the user does not add in the optional 8155-C chip. But MECH 1 leaves room for its initialization, just in case, and in the listing it is set up as three input ports. In the case of all C/S registers, the MVI instruction followed by an OUT instruction is the format for writing the set-up data into the registers. These steps run from 0008 to 0013 (hex) in RAM memory.

There are a few final things to initialize. For one thing, in our discussion on the interrupt system of the 8085A, we described what is called the interrupt stack. This is merely an area of RAM memory which the programmer can designate for the 8085A to use whenever an interrupt occurs, using those four special pins, TRAP, RST 5.5, RST 6.5, and RST 6.5, and RST 7.5. When the interrupt occurs, the 8085A automatically stores the 16-bit address of the next RAM address of its regular program into two consecutive locations in the stack. This way, when the interruption is over, the 8085A can retrieve that address from the stack, and return to its normal program exactly where it left off. But in order to designate the stack location, the programmer must load the 16-bit register in the 8085A which is termed the stack pointer. This register, (abbreviated SP) holds the 16-bit address of the highest location of the area of memory the user wishes to set aside as a stack. Thus, as in Mech 1, suppose the user wants to set aside the twenty RAM spaces from 00EC to 00FF (the last twenty spaces in 8155-A) as the Stack. He must load the 16-bit binary number for OOFF into the SP register; the 8085A will then use anything from OOFF down... and it should never have to use more than twenty address spaces, at least in MECH 1. After all, it uses only two spaces when an interrupt occurs, and there is one program in MECH 1 that uses the stack to store some other data as well, but is it all within the twenty-space margin of safety.

The LXI instruction takes its own second and third bytes and loads them into any selected register-pair in the 8085A, and SP is included as an option. Thus, in RAM locations 0014 through 0016 (hex) is an LXI instruction with the two selection bits set to indicate the loading of SPs. The second byte is 11111111, the highest RAM location in an 8155 chip, and the third byte is the address which specifies the 8155-A chip of the three possibilities.

The last initialization task is the set-up of the interrupt system itself. Figure 16-2 shows the operation of the SIM instruction. With SIM, it is possible to "mask," or make the 8085A unresponsive to, certain of the interrupt pins. In MECH 1, only two of the pins are used: TRAP and RST 5.5. (TRAP is unmaskable at any rate.) And so, we want to "mask" the other two RST pins. The byte to load, then, would be 00011110, which masks RST 6.5 and 7.5 and also resets the flip-flop associated with RST 7.5 (a feature we have not discussed and which need not concern us here). Bit 3 of the byte to load is set to a 1, which indicates that the byte is to be used to set the masks. If it were a 0, the masks would be unaffected, and the SOD pin of the 8085A (a function unused in the robot pet) could be set. The byte to load is placed into 8085A register A, and SIM moves it

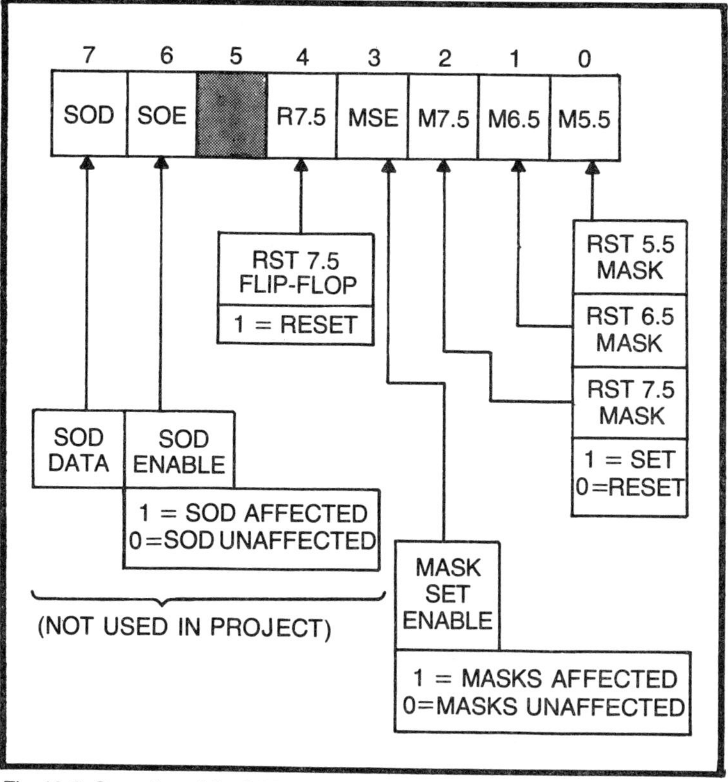

Fig. 16-2. Operation of the SIM instruction. Data is read from register A and used to set the indicated masks.

from there into the "masking" area of the interrupt system. Once this is done, an EI instruction (Enable Interrupt) is used to activate the interrupt system for use. RAM addresses 0017 to 001A (hex) contain the steps for interrupt initialization as just described.

With the completion of the initialization sequence, the program is ready to move on to more important matters. And so, a JMP instruction leaps the program flow to RAM address 400E (hex) where the random subroutine selection program termed ARASEM begins. Why jump all the way into 8155-B? Some more important program subroutines are to be dealt with, which will fill the remainder of 8155-A. These will be described next.

INTERRUPT ADDRESSES

When the *interrupt* pins are activated by some external device, the 8085A jumps to a certain address in RAM, keeping track of where it came from in the stack so that it can later return. TRAP jumps the program to 0024 (hex); RST 5.5 jumps it to 002C (hex); RST 6.5 jumps it to 0034 (hex); and RST 7.5 jumps it to 003C (hex). There are eight RAM spaces at each of these locations—not enough for much of a routine! But is is enough to place a JMP instruction to get to a larger area of RAM. These eight-space areas should be left for interrupt-related programming only, obviously.

At location 0024 (hex), then, should be the program (or the beginning of it) which handles the *battery monitor* function described in Chapter 12. When the battery monitor circuit senses that the batteries are low in charge, it signals the TRAP pin, which jumps the program to 0024 (hex). At this point, two things occur, according to our program listing. First a DI instruction disables the interrupt system, so that nothing can "interrupt the interruption," so to speak. Second, a JMP instruction leaps the program to address 0044 (hex), where the battery monitor routine is actually located.

On the other hand, suppose that you speak a Fredian word into the Excom circuit of the pet. The circuit, upon hearing the first note, would notify the RST 5.5 pin of the fact. The program would jump to 002C (hex), where a number of things occur. First is the DI instruction, as described above in the case of TRAP. Second is a series of PUSH instructions. The PUSH instructions actually takes all of the internal 8-bit registers of the 8085A (A, B, C, D, E, H, and L), plus the status of the internal flags, and stores these in the stack. As we'll see, if the Excom alert turned out to be a "false alarm," the Excom program would retrieve this data from the stack and would return to the original program as before the interruption. After the series of

PUSH instructions, a JMP instruction leads to address 0078, the location of the Excom routine.

Notice in the listings that the eight-byte areas for RST 6.5 and 7.5 are left blank in MECH 1. This is to permit the use of these interrupt pins at a latter date, should the user wish, though they are not presently used in the prototype of our robot dog. Remember, of course, that if these other pins are to be used, the data for the SIM instruction, located at 0018 (hex) would have to be altered accordingly.

In discussing these interrupt sections of memory, the block of addresses from 001E to 0022 has been passed over. These locations belong to the Excom subroutine, which will be described a bit later on.

BATTERY MONITOR ROUTINE

The first full-blown routine begins at 0044 (hex), and it is the routine which responds to a signal from the battery monitor circuit via the TRAP pin. To conserve space, not every instruction will be labored over, but major blocks will be explained.

Figure 16-3 gives a flowchart for the battery monitor subroutine. In essence, here is the outward result of the program. Upon sensing a weak battery, the pet ceases all motion, bringing its drive wheel to a centered stop. Then it begins to emit a series of individual barks. One bark occurs every four seconds or so, and each bark is different.

There will occur 16 different barks, which is the full range available from the Audigen circuit. After the train of barks, which takes about a minute to emit, the pet robot voluntarily deprograms its 8155 ports, rendering it paralyzed, and enters a programmed *halt*. Thus, if low supply voltage should cause the 8085A to act erratically, the chances of it jumping out of halt, reprogramming its ports and thus doing some damage are very slim. Once in this halt mode, there is only one way out...an external reset by the owner, who has recognized the train of barks as a hunger signal.

RAM locations 0044 to 0047 (hex) order the Servodrive to cease motion and to center the drive wheel. Steps 0048 to 004D (hex) set up the available *event timer* of 8155-A for a repeating 4-second delay, which can be looked for at port BA_2. The delay is set up by the summoning of a special event timer subroutine which will be described later. Steps 004E through 0058 (hex) set up register B as a counter of sorts, which can be moved into register A and sent to the Audigen ciruit at will. The JZ instruction at 0056 (hex) checks to see if the register has counted through all sixteen of the possible barks. Steps 0059 to 005E (hex) send the contents of B out to the Audigen and then quickly reset the Audigen to zero, which results in a single bark at a time. Steps 005F to 0065 (hex) set up a *waiting loop* which looks for the end of the repeated four-second delay. When the delay is up, the JMP instruction at 0066 (hex) begins the process over at the next bark. When all of the barks are completed, steps 0069 to 0071 (hex) "deconfigure" the 8155's and halt the processor.

Thus, in about fifty steps or so, the pet dog responds to a dangerous low-battery condition, not only with a shutdown, but with a distinctive signal to alert the owner to its plight. This demonstrates the power of the instruction set clearly. You may wish to use this routine or substitute something simpler, a basic shutdown. But the point is that the pet robot is versatile enough to do almost *anything* when the time comes. That is the beauty of software.

EXCOM ROUTINE

Remember that the pet canine system has two basic "pillars" upon which it rests its goal structure. One is the ARASEM concept, the free-choice selection of subroutines when the pet dog is not specifically under the command of its owner. The other is the Excom concept, in which external orders from the human owner supersede the personal choices of the pet. The Excom routine is somewhat involved, but that is to be expected, in view of what it does. As it is described in the following pages, resist the temptation to skip this section—the Excom routine is *central* to MECH 1, and much cannot be understood without it.

As we have seen, the presence of a *first reference pitch* of a Fredian word triggers the Excom circuitry, which signals the RCU-85 via RST 5.5. The program flow is jumped to location 002C (hex), where the 8085A registers are stored in the interrupt stack. Then the program jumps to location 0078 (hex), where the Excom routine begins in earnest.

Figure 16-4 shows the basic interactive flow of the routine. Essentially, the program waits for about three-tenths of a second, to allow the incoming frequency to stabilize, at which point it stores the frequency, which it reads as an 8-bit word at port AA, into a location in memory. It waits for the pitch to cease, indicating the presence of a Fredian *sync space*. It then initiates a five-second time-sequence; if no new frequency is received during this period, the first note or notes are assumed to have been a false reception. In such a case, the interruption is terminated, the registers are restored, and the original program is continued. If a new pitch comes in, though, the process repeats until all five Fredian pitches are stored.

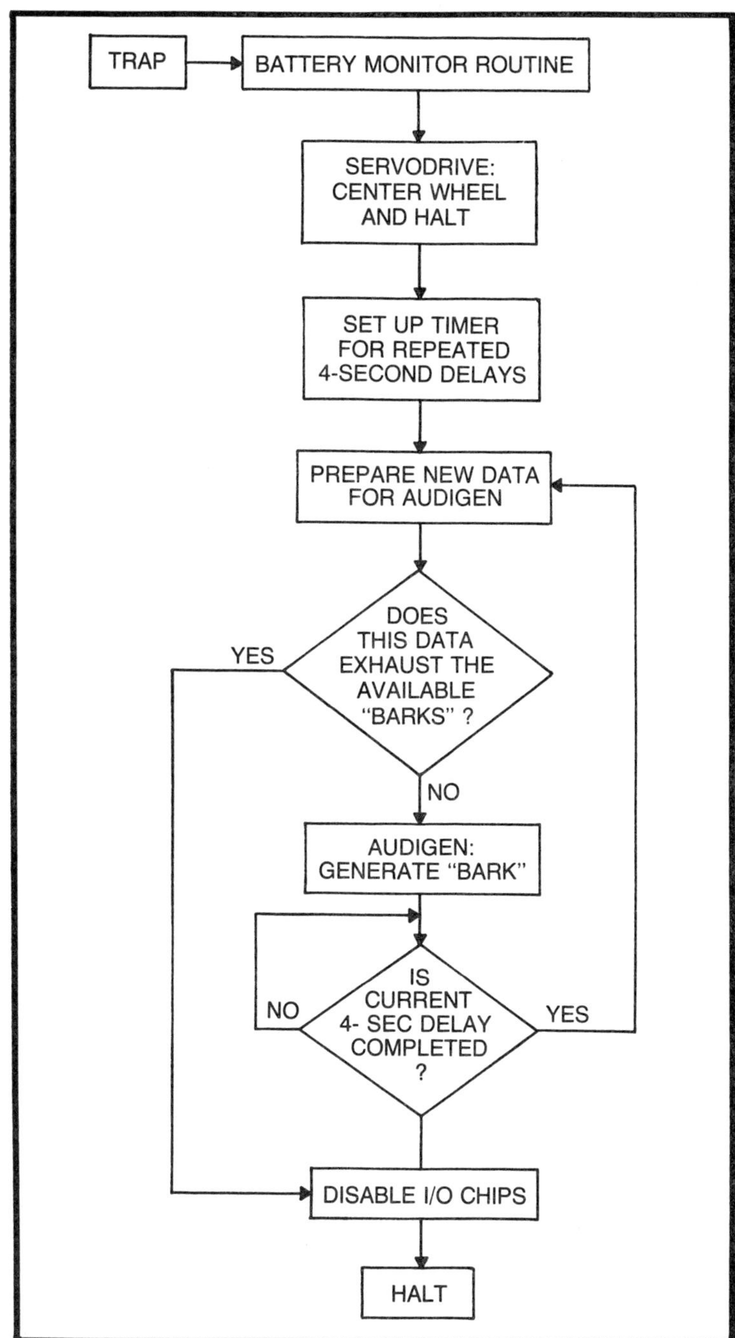

Fig. 16-3. Battery monitor routine.

Then the decoding process begins. The routine takes the first received pitch and compares it to each succeeding pitch, noting which is larger (higher in pitch). The result is a reassembled four-bit work, indicating one of 16 possible subroutines to be selected. The routine then uses this number to select one address out of a table of 16, which address it uses to jump to the requested subroutine. In MECH 1, to conserve space, the 16 subroutines available via Excom are the same 16 available to the ARASEM selector. When the subroutine is completed, therefore, it does *not* return to the place where it left off when the interruption came; rather, it returns to the ARASEM selector to find a new random subroutine.

Steps 0072 to 0074 load the register-pair D-E with the first address of a section of memory where the five pitches will be stored. These five locations are 001E through 0022 (hex), earlier in the 8155-A. The subsequent program, then, to store the next pitch, simply increments D-E, which would then point to the next of the five addresses.

Steps 0075 to 007A (hex) set up the event timer to delay for about three-tenths of a second before reading in the pitch, using the CALLable event timer subroutine. The reason is to give the Excom circuit time to obtain an accurate reading. Steps 007B to 0081 (hex) place the program into a loop until the 3/10-second delay is up. Then steps 0082 to 0085 (hex) load the pitch into the location specified by D-E, and increment D-E for the next pitch. Steps 0086 to 008B (hex) check to see if all five pitches have been loaded yet, by checking the status of D-E. If so, it jumps elsewhere.

If there are more pitches to come, steps 008C to 0092 wait for the sync space; then steps 0093 to 0098 (hex) start a five-second *validity period* using the CALLable event timer subroutine. Steps 0099 to 00A6 (hex) determine if a new pitch has come in during the validity period or not. If so, it jumps back to 0075 and repeats the whole loading process. If not, steps 00A7 to 00AC (hex) restore the 8085A registers and return the program flow to its original location before the Excom nterruption.

Once the five pitches have been loaded, the decoding occurs, in steps 00AD to 00BD (hex). This section of the routine steps through the last four locations where the pitches are stored, comparing those pitches to the first pitch stored, the reference pitch. The use of the CMP instruction sets the carry flag of the 8085A to a "1" if the pitch is higher than the reference, or to a "0" if it is lower. After each comparison, register B is moved into A, and a RAL instruction rotates the carry data into the lowest bit of the register. After four times, the decoded Fredian word is stored in register B.

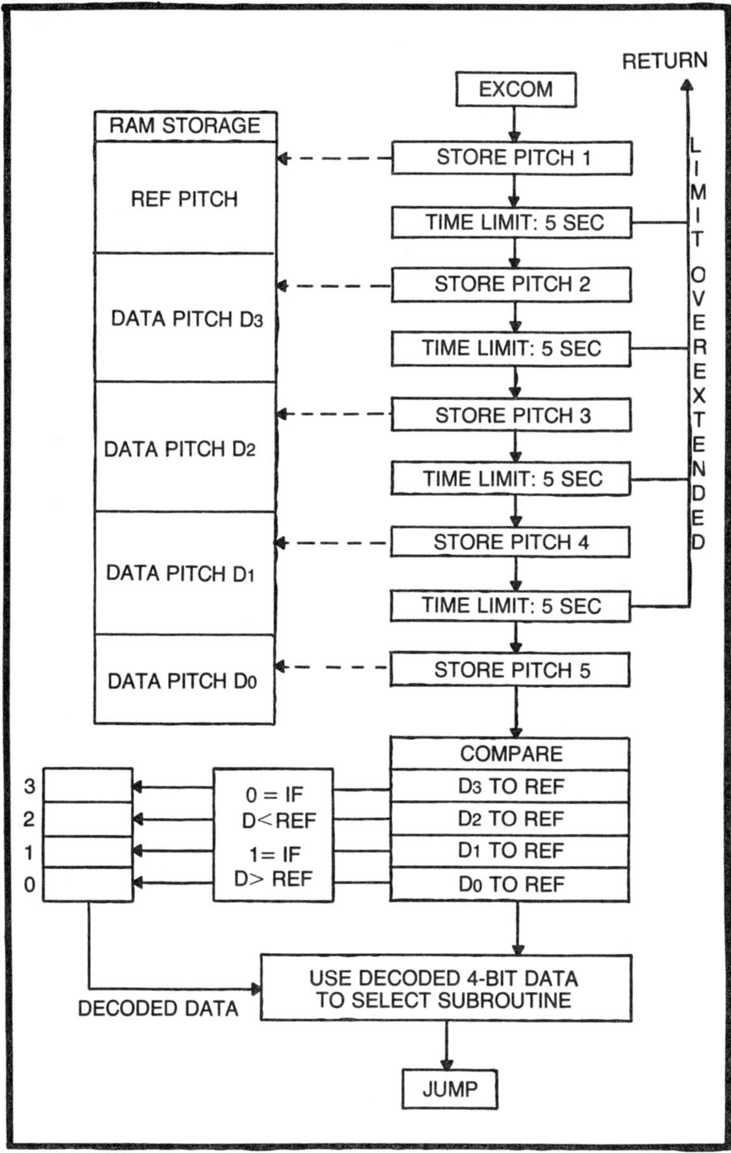

Fig. 16-4. Overview of the Excom storage and decoding processes.

The final step of the Excom routine is to use the four-bit Fredian data to select one of sixteen subroutines The 16-bit addresses for the subroutines are stored by the programmer in address locations 00CC to 00EB (hex), by placing the lower eight bits in one location and the upper eight bits in the next location. And so, steps

00BE to 00CB (hex) take the four-bit word, double it and add it to the first address of the subroutine table (00CC hex). In this way, each of 16 Fredian words can point to a 16-bit subroutine address in the table. Register-pair H-L is set up to point to the subroutine address, which is retrieved from the table into Pair D-E. The XCHG instruction moves this address to H-L, and the program jumps to the address and performs the subroutine. (The remaining 20 address locations in 8155-A, 00EC to 00FF, have been designated as the interrupt stock, and thus are left blank.)

The Excom routine takes up a good deal of the pet dog's memory space (as do most programs that interface a computer to the outside world), but its power is phenomenal. It enables a working, albeit simplified, vocal interface to occur between pet and owner. A more complicated version of this program could be evolved, allowing for a greater vocabulary, but such a newer version would merely be an expansion of the basic philosophy demonstrated in this routine.

ARASEM ROUTINE

The other important leg of the pet robot goal structure is the ARASEM concept. ARASEM stands for ARtificially RAndom SElf-Motivation, and is the process by which the robot pet chooses random courses of action when not under the direct Excom command of its owner. The fact that *both* Excom and ARASEM are welded into one interactive whole in MECH 1 is what enables the pet to simulate animal life to the high degree that it does.

The flowchart offered in Fig. 2-2, shows the program philosophy of the ARASEM process. For each of the 16 possible subroutines, there is a *probability number*, which the programmer assigns to it, stored in a special *probability table* in RAM. The ARASEM routine begins by generating a random number and using it to select one of the 16 possibilities as a *Candidate Subroutine*. Then a second random number is compared with the probability number of the candidate; if the random number is smaller, the subroutine is performed. If the random number is larger than the probability number, the ARASEM process begins all over again with a new candidate elected at random.

To perform this sort of activity, the program requires a good *random number generator* of some sort. Steps 4000 to 400D (hex) contain such a generator, which can be used by any program in MECH 1 simply by summoning it with a CALL instruction. Essentially, the generator takes any number, including zero, and multiplies it by 13 and adds 1. The lowest 8 bits of the number are saved at location 400D (hex), and are used as the basis of a next random number. Notice the RET instruction at location 400C (hex); regardless of where this subroutine was CALLed from, RET will return the program flow to the proper location, using the interrupt stack.

The ARASEM routine proper begins at 400E (hex), where a CALL instruction summons the production of a first random number. Steps 4011 to 401A (hex) use the four lowest bits of this number to select the candidate subroutine and to retrieve its probability number from a table listed later. Steps 401B to 4022 (hex) call forth a second random number and compare it to the probability number. A JC instruction starts the ARASEM process over from scratch if the comparison is not correct (if RND NO is greater than PROB NO).

If the comparison is favorable, however, steps 4023 to 4027 use the lower four bits of the first random number to look up the subroutine in the same subroutine table used by the Excom routine, in RAM locations 00CC to 00EB (hex). This is done actually by jumping to the very last portion of the Excom program in which this is done, in steps 00C2 to 00CB (hex).

The probability table is located, then, from 4028 to 4037 (hex). Each probabilitity number is an eight-bit word; the order of storage is the same as for the earlier subroutine table, corresponding as one probability number for each 16-bit address.

Now that both the ARASEM routine and the Excom program are instituted, it is possible to write in the subroutine programs, the addresses of which are stored in the subroutine table. If the pet robot uses only two 8155s, there is not much room for large, complicated subroutines, but even a few small ones are enough to make the pet into a versatile life-simulator. If 8155-C is added, though, there is much more freedom of space.

CALLABLE ROUTINES

Since space is at such a premium, it is a wise move to write any routines which may be used repeatedly into a CALLable routine, so that it need not be written more than once; the random number generator is a good example of this policy. There are two other such cases that would simplify things. One is the routine which operates the Soniscan system, and the other is the event timer.

Figure 16-5 shows the CALLable Soniscan monitor in flow chart form. The Soniscan circuit is loaded to aim its "beam" to a given direction; if this is the third and final direction, part of the monitor notes the fact. The monitor waits for an echo to return, with a validity period of about a tenth of a second as an upper waiting limit.

If an echo is received in that time, the monitor interrogates the Soniscan Counter in 8155-B and stores the count number. If no echo comes, the obstacle is assumed to be out of range, and the number "zero" is stored. The monitor stores three numbers, one for each direction, and then returns to the program which CALLed it. Each number is then available as a relative distance measurement; the higher the number, the closer the obstacle, with a range of from 00 to FF (hex).

This Soniscan monitor begins at location 4038 (hex). The first section, up to step 4043 (hex), begins by enabling a Soniscan transducer; then this case is checked to see if it is the last transducer or not. If all three transducers have been used, the program returns to the prior routine which requested the monitor.

If the job is not complete, though, steps 4044 to 4049 (hex) initiate a validity period of one-tenth of a second, by CALLing the event timer routine yet to be described. Then steps 404A to 4059 (hex) wait for the echo or the end of the time delay, whichever comes first. If the ehco comes first, steps 405A to 405C (hex) read the Soniscan counter and store the number; if the time delay runs out first, these steps are skipped, and "zero" is left in the location, having been stored in step 404A (hex). Then steps 405D to 4060 (hex) increment register pair H-L, which points to the storage location for the Soniscan numbers, and the process begins again. Locations 0061 to 0063 (hex) are for storage of the three numbers.

The Soniscan monitor is lengthy, and it is immediately visible that by making it a non-repeated CALLable subroutine, much space will be saved in the long run. After all, most of the general ARASEM or Excom subroutines which involve any motion of the pet ought to use Soniscan to prevent collision with obstacles. Continual repetition of this monitor would have been impractical.

A simpler CALLable subroutine is the event timer. All this routine does is load the two bytes stored in registers B and C respectively into the HI and LO registers of the event timer, and then issue the command to the C/S register to start the timer. Then the program flow returns to the prior routine which summoned the timer. Since the timer is sometimes used for a delay, and sometimes used for a time *limit*, this CALLable routine does not include steps to wait for the time to run out before returning. The routine takes the RAM area from location 0064 to 006E (hex). If a program wishes to CALL the timer, it uses an LXI instruction to load B and C with the bytes necessary for the desired timing, and then executes a CALL to the routine.

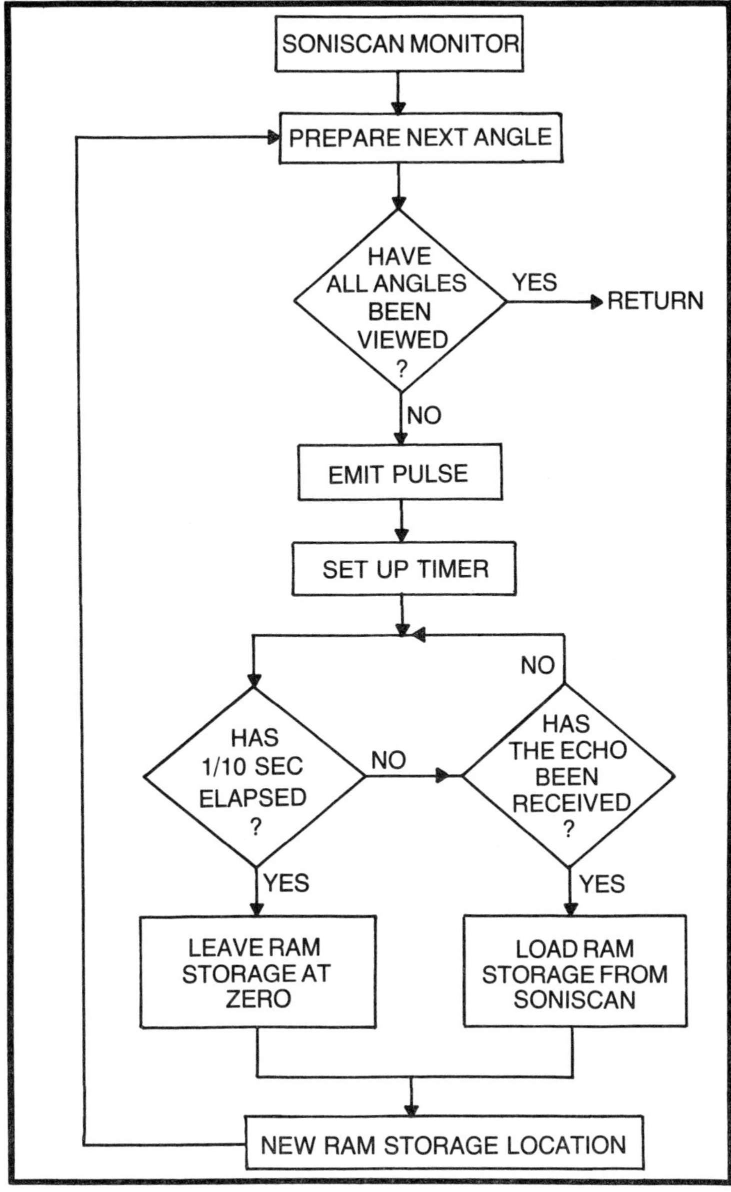

Fig. 16-5. Soniscan monitor routine.

GENERAL SUBROUTINES

The remainder of RAM memory, from 406F (hex) onward, may be used for Excom and ARASEM subroutines. These possible routines may range from simple to complex, depending on the amount of RAM space left; i.e., whether a third 8155 is used in the RCU-85. Even with the lesser amount of RAM, though, some interesting possibilities are open. A few such subroutines are listed in the MECH 1 table, but the reader is encouraged to write and experiment with his own.

One of these subroutines, termed the *watchdog* routine, is of particular note for those who might look at the mechanical pet and question its usefulness. The routine essentially scans the area immediately surrounding the pet. If the pet detects the presence of an obstacle any closer than a predetermined limit, it breaks into the Audigen equivalent of a "growl." The pet can thus act as a "night watch-dog" against such undesirable obstacles as burglars and stray cats. *Man's best friend*, indeed!

A few other sample subroutines are given in the listings. Note that the addresses for these have been placed in the subroutine table (beginning at 00CC), and that probability numbers have been arbitrarily assigned to them in the probability table at 4028. Note, too, that the first two examples are *loop* programs; if they are called by Excom, they can only be stopped by a reset, and if they are selected by ARASEM, they can only be exited by an Excom interruption or a reset. The second two examples are standard versions, which run for a given time and then return to ARASEM control. At the end of such a routine, a JMP instruction returns the program flow to address 0000 (hex), which leads the program into a re-initialization and a return to the ARASEM routine.

The examples given are simple programs, but they illustrate what is possible within the framework of the RCU-85 control structure. The system supports up to 16 subroutines, but to take full advantage of this range, it is best to add the third 8155 chip. As each new subroutine is written into RAM, its first address is stored in the subroutine table, lower address first, upper address second. The unused locations in the subroutine table should be left filled with zeroes; thus, if ARASEM selects from an unassigned place in the table, the program will merely jump to address 0000 (hex), and return for a different ARASEM selection. In addition, for each subroutine added, a number *must* be placed into the probability table for it. The larger the number, the greater the chance that it will be selected; thus, if there is a "zero" in the table location, the subroutine will never be chosen. This provides extra protection, too, if ARASEM selects an undesignated subroutine; if you leave a zero in its probability table location, it will never be performed.

FINAL PROGRAMMING NOTES

There are a few factors to remember in the writing of your own subroutines for the robot canine. First, pay strict attention to addressing. If you use a JMP instruction with an incorrect address, the RCU-85 will jump out into an area for which there is no RAM, and may react quite unpredictably. Remember, too, to keep track of which 8155 chip the program is written on. If a program must be continued on the next chip, DO NOT simply write it across the boundary. Use a JMP instruction to leap to a specific location in the next chip.

Second, pay strict attention to instructions. Don't try to run a program until you have written it down clearly on paper and have gone through it carefully several times. Reread the 8085A instruction set and get a feel for the options available to you.

Finally, make careful use of the two programming aids developed thus far: the *manual programmer* and the *tape interface*. Using the manual programmer, write the program into RAM one step at a time, using the hexadecimal equivalents of the 8-bit words and instructions. Then, hook up the tape interface, and use the *manual store* mode to read what you have just written, checking for errors. Once the program is proofread, activate the pet and prove out your program. Subject it to any parameters which might possibly crop up. Then, once you are certain of its workability, reset the canine robot and use the tape interface to store the whole of RAM memory onto a cassette tape. This stored program may not be your last or best, but at least it can be reloaded at will, using the interface, instead of reloading manually.

The pet dog project, as a prototype, is now completed and ready to be used to the fullest. MECH 1 endows the pet with an integrated goal structure that is both self-determining and obedient. The pet is also quite expandable, both in terms of hardware and software. Some will argue that the robot pet is not the "last word" in robotic endeavor; that its senses are limited and its circuitry simplified. This is true: our friend is *not* the *ultimate* robot pet. But it may well be called the *optimum* robot pet—possessing the many characteristics necessary for life simulation, yet in a simplified form to be practical and reproducible.

And yet, since our pet robot is *not* the "final word" in robot pets, the reader may have a burning desire to take the project still further in complexity. Thus, with a view to those whose time and resources permit, some engineering notes for a "second generation" pet are offered in the next chapter.

Table 16-1. MECH 1 Software Package (continued on pages 203-208).

ADDRESS	INSTRUCTION	MACHINE CODE	HEX CODE	EXPLANATION
0000	MVI$_{A,data}$	00111110	3E	
0001		00000000	00	
0002	OUT$_{TMR\text{-}LO\text{-}B}$	11010011	D3	
0003		01000100	44	
0004	MVI$_{A,data}$	00111110	3E	
0005		01000001	41	
0006	OUT$_{TMR\text{-}HI\text{-}B}$	11010011	D3	
0007		01000101	45	
0008	MVI$_{A,data}$	00111110	3E	
0009		00001100	0C	
000A	OUT$_{C/S\text{-}A}$	11010011	D3	
000B		00000000	00	
000C	MVI$_{A,data}$	00111110	3E	
000D		11001100	CC	
000E	OUT$_{C/S\text{-}B}$	11010011	D3	INITIALIZATION
000F		01000000	40	
0010	MVI$_{A,data}$	00111110	3E	
0011		00000000	00	
0012	OUT$_{C/S\text{-}C}$	11010011	D3	
0013		10000000	80	
0014	LXI$_{SP}$	00110001	31	
0015		11111111	FF	
0016		00000000	00	
0017	MVI$_{A,data}$	00111110	3E	
0018		00011110	1E	
0019	SIM	00110000	30	
001A	EI	11111011	FB	
001B	JMP$_{addr}$	11000011	C3	
001C		00001110	0E	
001D		01000000	40	
001E				
001F				
0020				EXCOM STORAGE (001E to 0022)
0021				
0022				
0023				
0024	DI	11110011	F3	
0025	JMP$_{addr}$	11000011	C3	
0026		01000100	44	
0027		00000000	00	TRAP SECTION
0028				
0029				
002A				
002B				
002C	DI	11110011	F3	
002D	PUSH$_{PSW}$	11110101	F5	
002E	PUSH$_{B\text{-}C}$	11000101	C5	
002F	PUSH$_{D\text{-}E}$	11010101	D5	
0030	PUSH$_{H\text{-}L}$	11100101	E5	RST 5.5 SECTION
0031	JMP$_{addr}$	11000011	C3	
0032		01110010	72	
0033		00000000	00	
0034				
0035				
0036				
0037				
0038				RST 6.5 SECTION
0039				
003A				
003B				
003C				
003D				
003E				RST 7.5 SECTION
003F				
0040				
0041				

Table 16-1 (continued)

ADDRESS	INSTRUCTION	MACHINE CODE	HEX CODE	EXPLANATION
0042				RST 7.5 SECTION
0043				
0044	MVI A,data	00111110	3E	
0045		00100000	20	
0046	OUT AC	11010011	D3	
0047		00000011	03	
0048	LXI B-C	00000001	01	
0049		00101000	28	
004A		11000000	C0	
004B	CALL TMR	11001101	CD	
004C		01100100	64	
004D		01000000	40	
004E	MVI B,data	00000110	06	
004F		00000000	00	
0050	MOV A,B	01111000	78	
0051	ADI data	11000110	C6	
0052		00000100	04	
0053	MOV B,A	01000111	47	
0054	CPI data	11111110	FE	
0055		01000000	40	
0056	JZ addr	11001010	CA	
0057		01101001	69	
0058		00000000	00	
0059	OUT BC	11010011	D3	BATTERY MONITOR ROUTINE
005A		01000011	43	
05B	MVI A,data	00111110	3E	
005C		00000000	00	
005D	OUT BC	11010011	D3	
005E		01000011	43	
005F	IN BA	11011011	DB	
0060		01000001	41	
0061	ANI data	11100110	E6	
0062		00000100	04	
0063	JNZ addr	11000010	C2	
0064		01011111	5F	
0065		00000000	00	
0066	JMP addr	11000011	C3	
0067		01010000	50	
0068		00000000	00	
0069	MVI A,data	00111110	3E	
006A		01000000	40	
006B	OUT C/S-A	11010011	D3	
006C		00000000	00	
006D	OUT C/S-B	11010011	D3	
006E		01000000	40	
006F	OUT C/S-B	11010011	D3	
0070		10000000	80	
0071	HLT	01110110	76	
0072	LXI D-E	00010001	11	
0073		00011110	1E	
0074		00000000	00	
0075	LXI B-C	00000001	01	
0076		00000100	04	
0077		10000000	80	
0078	CALL TMR	11001101	CD	
0079		01100100	64	
007A		01000000	40	EXCOM ROUTINE (Word Acquisition Section)
007B	IN BA	11011011	DB	
007C		01000001	41	
007D	ANI data	11100110	E6	
007E		00000100	04	
007F	JNZ addr	11000010	C2	
0080		01111011	7B	
0081		00000000	00	
0082	IN AA	11011011	DB	
0083		00000001	01	

Table 16-1 (continued)

ADDRESS	INSTRUCTION	MACHINE CODE	HEX CODE	EXPLANATION
0084	STAX$_{D-E}$	00010010	12	
0085	INX$_{D-E}$	00010011	13	
0086	MOV$_{A,E}$	01111011	7B	
0087	CPI$_{data}$	11111110	FE	
0088		00100011	23	
0089	JZ$_{addr}$	11001010	CA	
008A		10101101	AD	
008B		00000000	00	
008C	IN$_{BA}$	11011011	DB	
008D		01000001	41	
008E	ANI$_{data}$	11100110	E6	
008F		00000010	02	
0090	JNZ$_{addr}$	11000010	C2	
0091		10001100	8C	
0092		00000000	00	
0093	LXI$_{B-C}$	00000001	01	
0094		00110010	32	
0095		10000000	80	EXCOM ROUTINE
0096	CALL$_{TMR}$	11001101	CD	—(Word Acquisition
0097		01100100	64	Section)
0098		01000000	40	
0099	IN$_{BA}$	11011011	DB	
009A		01000001	41	
009B	ANI$_{data}$	11100110	E6	
009C		00000110	03	
009D	CPI$_{data}$	11111110	FE	
009E		00000100	04	
009F	JZ$_{addr}$	11001010	CA	
00A0		10101011	AB	
00A1		00000000	00	
00A2	CPI$_{data}$	11111110	FE	
00A3		00000110	03	
00A4	JZ$_{addr}$	11001010	CA	
00A5		01111011	7B	
00A6		00000000	00	
00A7	POP$_{H-L}$	11100001	E1	
00A8	POP$_{D-E}$	11010001	D1	
00A9	POP$_{B-C}$	11000001	C1	
00AA	POP$_{PSW}$	11110001	F1	
00AB	EI	11111011	FB	
00AC	RET	11001001	C9	
00AD	LXI$_{H-L}$	00100001	21	
00AE		00011110	1E	
00AF		00000000	00	
00B0	LDA$_{addr}$	00111010	3A	
00B1		00011110	1E	
00B2		00000000	00	
00B3	INX$_{H-L}$	00100011	23	
00B4	CMP$_{M}$	10111110	BE	
00B5	MOV$_{A,B}$	01111000	78	
00B6	RAL	00010111	17	
00B7	MOV$_{B,A}$	01000111	47	
00B8	MOV$_{A,L}$	01111101	7D	EXCOM ROUTINE
00B9	CPI$_{data}$	11111110	FE	–(Data Decoding
00BA		00100010	22	Section)
00BB	JNZ$_{addr}$	11000010	C2	
00BC		10110000	B0	
00BD		00000000	00	
00BE	MOV$_{A,B}$	01111000	78	
00BF	ANI$_{data}$	11100110	E6	
00C0		00001111	0F	
00C1	ADD$_{A}$	10000111	87	
00C2	LXI$_{H-L}$	00100001	21	
00C3		11001100	CC	
00C4		00000000	00	
00C5		10000101	85	

Table 16-1 (continued)

ADDRESS	INSTRUCTION	MACHINE CODE	HEX CODE	EXPLANATION
00C6	MOV L,A	01101111	6F	EXCOM ROUTINE (Data Decoding Section)
00C7	MOV E,M	01011110	5E	
00C8	INX H-L	00100011	23	
00C9	MOV D,M	01010110	53	
00CA	XCHG	11101011	EB	
00CB	PCHL	11101001	E9	
00CC		(01101111)	SUB0ADDR (406F)	SUBROUTINE TABLE
00CD		(01000000)		
00CE		(10001100)	SUB1ADDR (408C)	
00CF		((01000000)		
00D0		(10111010)	SUB2ADDR (40BA)	
00D1		(01000000)		
00D2		(11000111)	SUB3ADDR (40C7)	
00D3		(01000000)		
00D4			SUB4ADDR	
00D5				
00D6			SUB5ADDR	
00D7				
00D8			SUB6ADDR	
00D9				
00DA			SUB7ADDR	
00DB				
00DC			SUB8ADDR	
00DD				
00DE			SUB9ADDR	
00DF				
00E0			SUBAADDR	
00E1				
00E2			SUBBADDR	
00E3				
00E4			SUBCADDR	
00E4				
00E5			SUBDADDR	
00E6				
00E7			SUBEADDR	
00E8				
00E9			SUBFADDR	
00EA				
00EB				
00EC to 00FF—Reserved as Interrupt Stack				
4000	LXI H-L	00100001	21	RANDOM NUMBER GENERATOR
4001		00001101	0D	
4002		01000000	40	
4003	MOV A,M	01111110	7E	
4004	ADD A	10000111	87	
4005	ADD A	10000111	87	
4006	MOV C,A	01001111	4F	
4007	ADD A	10000111	87	
4008	ADD C	10000001	81	
4009	ADD M	10000110	86	
400A	INR A	00111100	3C	
400B	MOV M,A	01110111	77	
400C	RET	11001001	C9	
400D	RANDOM NUMBER STORAGE LOCATION			
400E	CALL RND	11001101	CD	ARASEM ROUTINE
400F		00000000	00	
4010		01000000	40	
4011	MOV A,M	01111110	7E	
4012	ANI data	11100110	E6	
4013		00001111	0F	
4014	MOV E,A	01011111	5F	
4015	LXI H-L	00100001	21	
4016		00101000	28	
4017		01000000	40	

Table 16-1 (continued)

ADDRESS	INSTRUCTION	MACHINE CODE	HEX CODE	EXPLANATION
4018	ADD$_{L}$	10000101	85	
4019	MOV$_{L,A}$	01101111	6F	
401A	MOV$_{D,M}$	01010110	56	
401B	CALL$_{RND}$	11001101	CD	
401C		00000000	00	
401D		01000000	40	
401E	MOV$_{A,D}$	01111010	7A	
401F	CMP$_{M}$	10111110	BE	
4020	JC$_{addr}$	11011010	DA	ARASEM ROUTINE
4021		00001110	0E	
4022		01000000	40	
4023	MOV$_{A,E}$	01111011	7B	
4024	ADD$_{A}$	10000111	87	
4025	JMP$_{addr}$	11000011	C3	
4026		11010100	D4	
4027		00000000	00	
4028	PROB$_{0}$	(11111111)	(FF)	
4029	PROB$_{1}$	(11111111)	(FF)	
402A	PROB$_{2}$	(11111111)	(FF)	
402B	PROB$_{3}$	(11111111`	(FF)	
402C	PROB$_{4}$			
402D	PROB$_{5}$			
402E	PROB$_{6}$			
402F	PROB$_{7}$			PROBABILITY TABLE
4030	PROB$_{8}$			
4031	PROB$_{9}$			
4032	PROB$_{A}$			
4033	PROB$_{B}$			
4034	PROB$_{C}$			
4035	PROB$_{D}$			
4036	PROB$_{E}$			
4037	PROB$_{F}$			
4038	LXI$_{H-L}$	00100001	21	
4039		01100001	61	
403A		01000000	40	
403B	MVI$_{D,data}$	00010110	16	
403C		00000000	00	
403D	INR$_{D}$	00010100	14	
403E	MOV$_{A,D}$	01111010	7A	
403F	CPI$_{data}$	11111110	FE	
4040		00000100	04	
401	RZ	11001000	C8	
4042	OUT$_{BC}$	11010011	D3	
4043		01000011	43	
4044		00000001	01	
4045	LXI$_{B-C}$	00000010	02	
4046		10000000	80	
4047		11001101	CD	
4048	CALL$_{TMR}$	01100100	64	SONISCAN ROUTINE
4049		01000000	40	
404A		00110110	36	
404B	MVI$_{M,data}$	00000000	00	
404C		11011011	DB	
404D	IN$_{BA}$	01000001	41	
404E	ANI$_{data}$	11100110	E6	
404F		00000101	05	
4050	CPI$_{data}$	11111110	FE	
4051		00000101	05	
4052	JZ$_{addr}$	11001010	CA	
4053		01001100	4C	
4054		01000000	40	
4055	CPI$_{data}$	11111110	FE	
4056		00000100	04	
4057	JNZ$_{addr}$	11000010	C2	
4058		01011101	5D	
4059		01000000	40	
405A	IN$_{TMR-LO-B}$	11011011	DB	

Table 16-1 (continued)

ADDRESS	INSTRUCTION	MACHINE CODE	HEX CODE	EXPLANATION
405B		01000100	44	
405C	MOV$_{M,A}$	01110111	77	
405D	INX$_{H,L}$	00100011	23	SONISCAN
405E	JMP$_{addr}$	11000011	C3	ROUTINE
405F		00111101	3D	
4060		01000000	40	
4061	SONISCAN LEFT DISTANCE			
4062	SONISCAN FRONT DISTANCE			
4063	SONISCAN RIGHT DISTANCE			
4064	MOV$_{A,B}$	01111000	78	
4065	OUT$_{TMR\text{-}HI\text{-}A}$	11010011	D3	
4066		00000101	05	
4067	MOV$_{A,C}$	01111001	79	
4068	OUT$_{TMR\text{-}LO\text{-}A}$	11010011	D3	
4069		00000100	04	EVENT TIMER
406A	MVI$_{A,data}$	00111110	3E	ROUTINE
406B		11001100	CC	
406C	OUT$_{C/S\text{-}A}$	11010011	D3	
406D		00000000	00	
406E	RET	11001001	C9	
SUBROUTINE STORAGE—406F to 40FF or to 80FF				
EXAMPLES FOLLOW				
406F	CALL$_{SON}$	11001101	CD	
4070		00111000	38	
4071		01000000	40	
4072	MVI$_{A,data}$	01111110	7E	
4073		11101011	EB	
4074	LXI$_{H\text{-}L}$	00100001	21	
4075		01100001	61	
4076		01000000	40	
4077	CMP$_{M}$	10111110	BE	SUB$_0$:
4078	JC$_{addr}$	11011010	DA	WATCHDOG ROUTINE-
4079		10000101	85	pet robot "barks"
407A		01000000	40	at anything that
407B	MOV$_{A,L}$	01111101	7D	approaches closer
407C	CPI$_{data}$	11111110	FE	than about ⅛foot:
407D		01100011	63	distance is in-
407E	JZ$_{addr}$	11001010	CA	creased by decreas-
407F		01101111	6F	ing the number in
4080		01000000	40	Location 4073;
4081	INX$_{H\text{-}L}$	00100011	23	exitable only by
4082	JMP$_{addr}$	11000011	C3	Reset or, if sel-
4083		01110111	77	ected by ARASEM,
4084		01000000	40	by Excom.
4085	MVI$_{A,data}$	00111110	3E	
4086		00001000	08	
4087	OUT$_{BC}$	11010011	D3	
4088		01000011	43	
4089	JMP$_{addr}$	11000011	C3	
408A		01101111	6F	
408B		01000000	40	
408C	CALL$_{addr}$	11001101	CD	
408D		10110001	B1	
408E		01000000	40	
408F	JNC$_{addr}$	11010010	D2	
4090		10011001	99	
4091		01000000	40	
4092	MVI$_{A,data}$	00111110	3E	
4093		00100001	21	
4094	OUT$_{AC}$	11010011	D3	
4095		00000011	03	
4096	JMP$_{addr}$	11000011	C3	
4097		10001100	8C	

Table 16-1 (continued)

ADDRESS	INSTRUCTION	MACHINE CODE	HEX CODE	EXPLANATION
4098		01000000	40	
4099	MVIA,data	00111110	3E	
409A		00000100	04	
409B	OUTAC	11010011	D3	
409C		00000011	03	
409D	INBA	11011011	DB	
409E		01000001	41	
409F	ANIdata	11100110	E6	
40A0		00001000	08	
40A1	JZaddr	11001010	CA	
40A2		10001101	8D	SUB1: TRAVEL ROUTINE—robot pet moves forward until it perceives an obstacle; it then stops, turns left, and proceeds until the obstacle is avoided; forward motion is resumed; exitable only by Reset or, if selected by ARASEM, by Excom.
40A3		01000000	40	
40A4	MVIA,data	00111110	3E	
40A5		00000101	05	
40A6	OUTAC	11010011	D3	
40A7		00000011	03	
40A8	CALLaddr	11001101	CD	
40A9		10110001	B1	
40AA		01000000	40	
40AB	JNCaddr	11010010	D2	
40AC		10101000	A8	
40AD		01000000	40	
40AE	JMPaddr	11000011	C3	
40AF		10010010	92	
40B0		01000000	40	
40B1	CALLSON	11001101	CD	
40B2		00111000	38	
40B3		01000000	40	
40B4	LDAaddr	00111010	3A	
40B5		01100010	62	
40B6		01000000	40	
40B7	CPIdata	11111110	FE	
40B8		11110000	F0	
40B9	RET	11001001	C9	
40BA	CALLRND	11001101	CD	
40BB		00000000	00	SUB2: BARK ROUTINE—pet robot emits one random "bark" and returns to ARASEM control.
40BC		01000000	40	
40BD	MOVA,M	01111110	7E	
40BE	OUTBC	11010011	D3	
40BF		01000011	43	
40C0	MVIA,data	00111110	3E	
40C1		00000000	00	
40C2	OUTBC	11010011	D3	
40C3		01000011	43	
40C4	JMPaddr	11000011	C3	
40C5		00000000	00	
40C6		00000000	00	
	LXIB-C	00000001	01	
		10010110	96	
		10000000	80	
	CALLTMR	11001101	CD	
		01100100	64	
		01000000	40	SUB3: CIRCLE ROUTINE—pet robot moves in a right-hand circle for 15 seconds, then returns to ARASEM control.
	MVIA,data	00111110	3E	
		00111001	39	
	OUTAC	11010011	D3	
		00000011	03	
40D1	INBA	11011011	DB	
40D2		01000001	41	
40D3	ANIdata	11100110	E6	
40D4		00000100	04	
40D5	JNZaddr	11000010	C2	
40D6		11001101	CD	
40D7		01000000	40	
40D8	JMPaddr	11000011	C3	
40D9		00000000	00	
40DA		00000000	00	

Chapter 17
Where From Here?

The field of robotics technology is by no means a conquered science, and the addition of our robot pet, though a good step forward, is only a step. All of the basic design problems in robotics come to a "head" in this project, and are resolved by simple, optimized approaches. This is satisfying, but for the avid experimenter, the satisfaction is but temporary. Cannot the rudimentary systems of the pet robot be brought forth to their logical conclusion?

Let us then consider those areas of the completed robot pet system which can be enhanced or extended to result in a high-level, "second generation" pet. Many of the following suggestions will require careful study and development, as well as further resources, to pursue. But all of these ideas are practical and attainable, in light of presently available circuitry and technology.

TOWARD BETTER LOCOMOTION

Chapter 3 noted the fact that, for locomotion of the robot pet, wheels were simpler to implement than legs. And yet, wheels do have their drawbacks, in terms of traction and the negotiation of certain terrain. And, after all, no animals observed to date make use of wheels; they could not survive if they did! A step toward a legged pet robot, if only for aesthetic reasons, is desirable.

Figure 17-1 shows a suggested arrangement for the use of legs in a second-generation pet. The side view only shows two of four legs, but the other side would operate identically, though the timing would be inverted.

Basically, what occurs is this. The front leg (F) and the rear leg (R) always move either toward each other or away from each other, never in phase. In addition to lateral motion, each leg can be lifted vertically, independently. The timing waveforms of Fig. 17-1 illustrate the cadence of the legs on a given side. As leg F is lifted, F and R move further apart; leg R is in contact with the ground and thrusts the body forward. Then leg F drops and R raises, and the legs move together; F hugs the ground and thrusts the body forward. The cadence on the opposite side is 90 degrees out of phase, as it were, so that when two legs are "up," they are *never* both front or rear legs. In this way, the pet is able to "get its feet off of the ground."

How would the pet be able to turn? The center of the body would, in actuality, have to be pivoted. An activating gearmotor would slightly bend the length of the body, which to a greater or lesser degree would result in a curved pathway for the pet. The motion would be erratic and awkward in many respects, and the robot pet might need shock absorbers, but it would have honest, functional legs. The idea is feasible enough to be worth considering.

TOWARD IMPROVED MEMORY

As Chapter 14 explained, the construction of a tape interface for the canine robot solves the problem of RAM in the RCU-85. RAM is volatile—it goes "blank" without supply voltage—but the Interface allows quick restoration of memory, just in case. And yet, the tape interface *is* an involved project! If there were another way, it would bear investigation.

The "other way" is the use of EPROM memory, which does *not* blank out without power. The drawbacks are two in number: 1) The obtaining of an EPROM programmer, and 2) The obtaining of an ultraviolet source to erase the EPROM. These two pieces of equipment can be quite expensive, which is why the all-RAM philosophy won out in this prototype. But if the reader wishes to use some EPROM, what should he do?

For any small system based on the 8085A, the best EPROM chip to use is the 8755 circuit. Figure 17-2 gives the block diagram for this useful device. The 8755 contains 2048 bytes of EPROM memory, plus two programmable I/O ports. Like the 8155 chip, the 8755 is fully compatible with the 8085A, having the characteristic AD and A lines, $\mathrm{IO}/\overline{\mathrm{M}}$, $\overline{\mathrm{RD}}$, ALE, and RESET inputs. A pet robot system with only one 8155 and one 8755 would have 256 bytes of RAM for such functions that require RAM (such as the Interrupt Stack), five programmable I/O ports, and 2048 bytes of EPROM for actual subroutines.

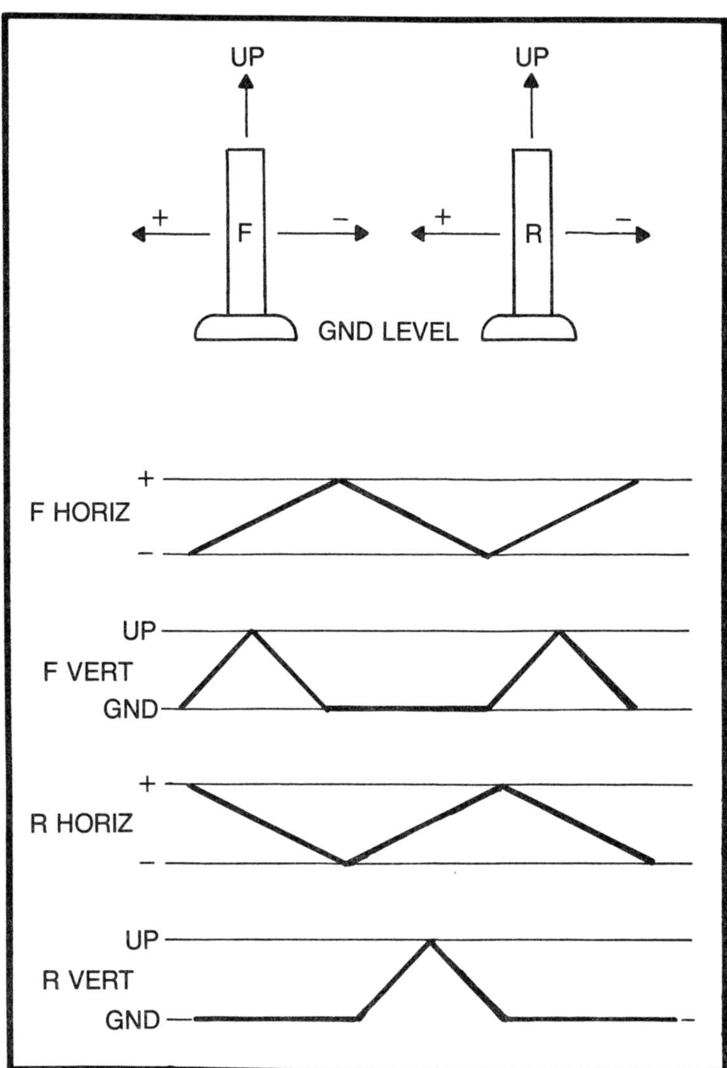

Fig. 17-1. A proposed system of locomotion using legs.

The 8755, in addition, is simple to program in comparison to other, more standard EPROM chips. A light +26-volt supply is run to the V_{DD} pin. Then the address is loaded with ALE, and the data present on the AD lines is loaded by a 50 millisecond low-to-high TTL pulse on the PROG/$\overline{CE}$ pin. Such a programmer would not be hard to design and construct.

As for erasing the 8755, the reader's best bet is the small UVS-11E, a small-volume ultraviolet EPROM eraser produced by

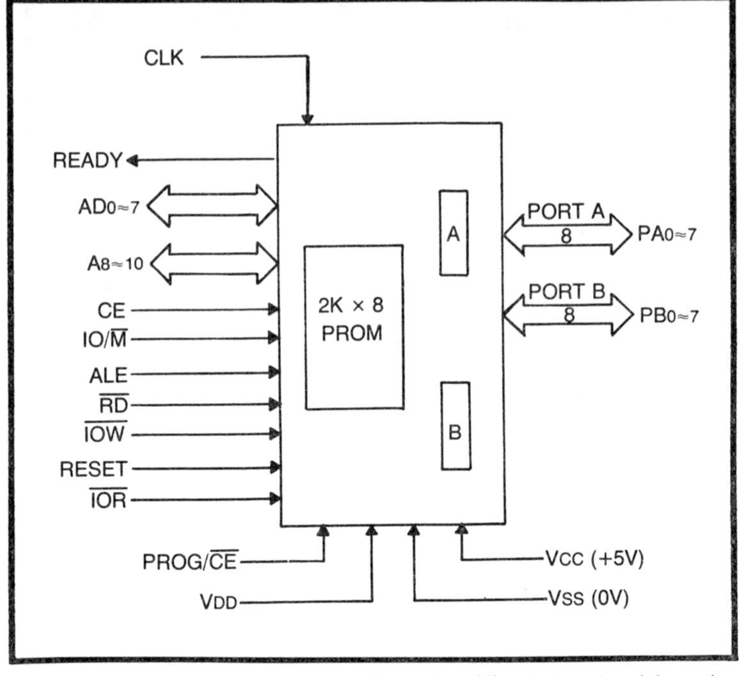

Fig. 17-2. Block diagram of the 8755 EPROM-I/O chip (reprinted from the MCS-85 User's Manual, copyright 1978, courtesy of Intel Corporation).

Ultra-Violet Products of San Gabriel, CA. This eraser is aimed at the hobbyist market, and is therefore produced quite reasonably.

The use of the 8755 is still a bit more costly than the all-RAM approach. But the experimenter who can afford to go this route should seriously consider it. The expanded memory capacity alone is enough to make this option an attractive basis for a second-generation pet dog.

TOWARD ENHANCED SOFTWARE

The present size of robot pet memory limits the number of possible subroutines to but a few. Fortunately, these are plenty to fully convey the life-simulation we are seeking. But if the memory capacity were to be expanded, say, by the addition of an 8755 to the RCU-85, the enhancements to software could be manifold.

One possible enhancement would be a wider application of the Excom subsystem and the attendant Fredian grammar. Presently, the Fredian word conveys only a four-bit word, to select one of 16 subroutines. With expanded memory, though, it would be possible to have at least 64 subroutines—possibly more. And so, the Fredian

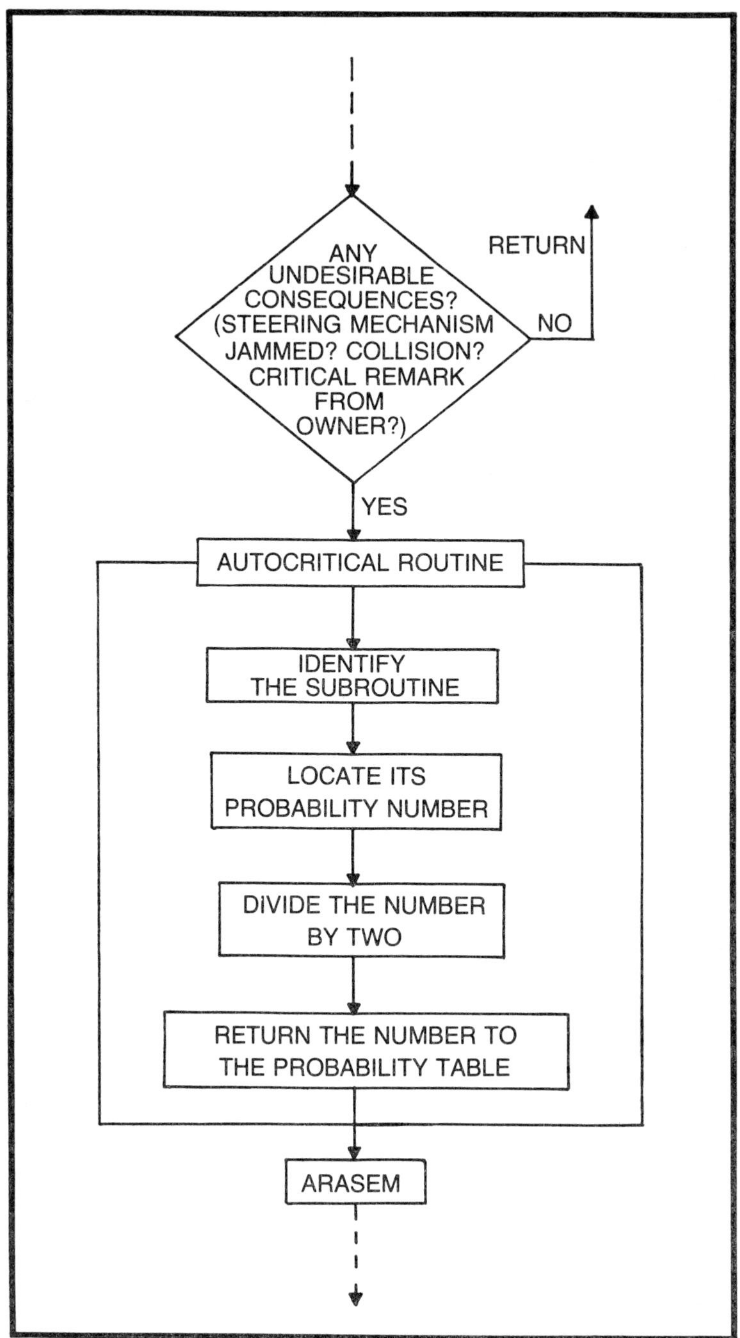

Fig. 17-3. Autocritical software flowchart.

word could be expanded to convey a six-bit word, thus expanding the pet's working "vocabulary."

Another even more significant possibility opened up by expanded memory is the implementation of self-modifying software. The rudiments of such an advance are present even now in the robot pet, with the presence of the probability table. But presently, these probability numbers are static, fixed entities, preset by the owner and shaped as he sees fit. On the analogy of life, though, it would be a greater achievement to write some sort of judging criteria into the RCU-85 software. Then the software could judge the advisability of this or that subroutine in a trial-and-error fashion, and adjust the probability numbers accordingly.

Figure 17-3 shows the flowchart for such programming, which may be termed *autocritical software*, in that it is "self-judging." For example, the pet robot finds that wherever it executes a subroutine which calls for extreme turning angles, it is more apt to "jam up" (which it might sense by noting an excessive current drain at the motor). This condition is deemed undesirable by pre-programmed standards—and so the autocritical routine *decrements* the probability number for that subroutine by several degrees, thereby effectively curtailing the chances that such jam-ups will occur in the future. True, it is the owner who has determined what is to be considered undesirable, but it is the RCU-85 that attempts to judge and correct its own actions to meet these standards. The robot canine becomes a *learning* pet.

A ROBOT PET IN EVERY HOME?

Wherever the ongoing development of the robot pet project may lead, one overarching keynote is *practicality*. In much of popular thought on robots today, there is a leaning toward the complex and ideal. The public expects full-blown electronic housemaids and butlers to come about with ease, not realizing the immense difficulties in getting a robot to keep from running into walls, much less getting it to wash dishes.

This volume, though, if nothing else, indicates that if robots are to become practical, we must aim first for things that a robot can do *well*. It may be years before a robot will be able to serve dinner or vacuum the floors on its own. But *today*, a robot can wag its tail and bark. A mechanical butler in every home is a distant possibility; but a robot pet in every home is a much closer and much more realistic goal.

And if you *still* doubt that this is true, then tell me: Just what *is* that unusual sound I hear coming from your workshop?

Appendices

Appendix A
Possible Motor Sources for the Robot Pet

American Bosch Electrical Products Division
McCrary Road
Columbus, Miss. 39701

Brevel Motors
203 Broad Street
Carlstadt, N.J. 07072

Inland Motor Division of Kollmorgen Corporation
501 First Street
Radford, VA. 24141

Mamco Corporation
532-542 Fourth Street
Racine, Wis. 53404

Motor Products—Owosso Corporation
201 S. Delaney Road
Owosso, Mich. 48867

Pittman Corporation
Harleysville, Penna. 19438

Transicoil, Inc.
Worcester, Penna. 19490

Voorlas Manufacturing Company
1711 South Street
Racine, Wis. 53404

Appendix B
Powers of Two

n	2^n	2^n THREE BYTE BINARY		
0	1	00000000	00000000	00000001
1	2	00000000	00000000	00000010
2	4	00000000	00000000	00000100
3	8	00000000	00000000	00001000
4	16	00000000	00000000	00010000
5	32	00000000	00000000	00100000
6	64	00000000	00000000	01000000
7	128	00000000	00000000	10000000
8	256	00000000	00000001	00000000
9	512	00000000	00000010	00000000
10	1024	00000000	00000100	00000000
11	2048	00000000	00001000	00000000
12	4096	00000000	00010000	00000000
13	8192	00000000	00100000	00000000
14	16384	00000000	01000000	00000000
15	32768	00000000	10000000	00000000
16	65536	00000001	00000000	00000000
17	131072	00000010	00000000	00000000
18	262144	00000100	00000000	00000000
19	524288	00001000	00000000	00000000
20	1048576	00010000	00000000	00000000
21	2097152	00100000	00000000	00000000
22	4194304	01000000	00000000	00000000
23	8388608	10000000	00000000	00000000

Appendix C
Hexadecimal-Decimal Integer Conversion

The following table provides for direct conversions between hexadecimal integers in the range 0–FFF and decimal integers in the range 0–4095. For conversion of larger integers, the table values may be added to the following figures:

Hexadecimal	Decimal	Hexadecimal	Decimal
01 000	4 096	20 000	131 072
02 000	8 192	30 000	196 608
03 000	12 288	40 000	262 144
04 000	16 384	50 000	327 680
05 000	20 480	60 000	393 216
06 000	24 576	70 000	458 752
07 000	28 672	80 000	524 288
08 000	32 768	90 000	589 824
09 000	36 864	A0 000	655 360
0A 000	40 960	B0 000	720 896
0B 000	45 056	C0 000	786 432
0C 000	49 152	D0 000	851 968
0D 000	53 248	E0 000	917 504
0E 000	57 344	F0 000	983 040
0F 000	61 440	100 000	1 048 576
10 000	65 536	200 000	2 097 152
11 000	69 632	300 000	3 145 728
12 000	73 728	400 000	4 194 304
13 000	77 824	500 000	5 242 880
14 000	81 920	600 000	6 291 456
15 000	86 016	700 000	7 340 032
16 000	90 112	800 000	8 388 608
17 000	94 208	900 000	9 437 184
18 000	98 304	A00 000	10 485 760
19 000	102 400	B00 000	11 534 336
1A 000	106 496	C00 000	12 582 912
1B 000	110 592	D00 000	13 631 488
1C 000	114 688	E00 000	14 680 064
1D 000	118 784	F00 000	15 728 640
1E 000	122 880	1 000 000	16 777 216
1F 000	126 976	2 000 000	33 554 432

HEXADECIMAL-DECIMAL INTEGER CONVERSION

	0	1	2	3	4	5	6	7	8	9	A	B	C	D	E	F
000	0000	0001	0002	0003	0004	0005	0006	0007	0008	0009	0010	0011	0012	0013	0014	0015
010	0016	0017	0018	0019	0020	0021	0022	0023	0024	0025	0026	0027	0028	0029	0030	0031
020	0032	0033	0034	0035	0036	0037	0038	0039	0040	0041	0042	0043	0044	0045	0046	0047
030	0048	0049	0050	0051	0052	0053	0054	0055	0056	0057	0058	0059	0060	0061	0062	0063
040	0064	0065	0066	0067	0068	0069	0070	0071	0072	0073	0074	0075	0076	0077	0078	0079
050	0080	0081	0082	0083	0084	0085	0086	0087	0088	0089	0090	0091	0092	0093	0094	0095
060	0096	0097	0098	0099	0100	0101	0102	0103	0104	0105	0106	0107	0108	0109	0110	0111
070	0112	0113	0114	0115	0116	0117	0118	0119	0120	0121	0122	0123	0124	0125	0126	0127
080	0128	0129	0130	0131	0132	0133	0134	0135	0136	0137	0138	0139	0140	0141	0142	0143
090	0144	0145	0146	0147	0148	0149	0150	0151	0152	0153	0154	0155	0156	0157	0158	0159
0A0	0160	0161	0162	0163	0164	0165	0166	0167	0168	0169	0170	0171	0172	0173	0174	0175
0B0	0176	0177	0178	0179	0180	0181	0182	0183	0184	0185	0186	0187	0188	0189	0190	0191
0C0	0192	0193	0194	0195	0196	0197	0198	0199	0200	0201	0202	0203	0204	0205	0206	0207
0D0	0208	0209	0210	0211	0212	0213	0214	0215	0216	0217	0218	0219	0220	0221	0222	0223
0E0	0224	0225	0226	0227	0228	0229	0230	0231	0232	0233	0234	0235	0236	0237	0238	0239
0F0	0240	0241	0242	0243	0244	0245	0246	0247	0248	0249	0250	0251	0252	0253	0254	0255

100	0256	0257	0258	0259	0260	0261	0262	0263	0264	0265	0266	0267	0268	0269	0270	0271
110	0272	0273	0274	0275	0276	0277	0278	0279	0280	0281	0282	0283	0284	0285	0286	0287
120	0288	0289	0290	0291	0292	0293	0294	0295	0296	0297	0298	0299	0300	0301	0302	0303
130	0304	0305	0306	0307	0308	0309	0310	0311	0312	0313	0314	0315	0316	0317	0318	0319
140	0320	0321	0322	0323	0324	0325	0326	0327	0328	0329	0330	0331	0332	0333	0334	0335
150	0336	0337	0338	0339	0340	0341	0342	0343	0344	0345	0346	0347	0348	0349	0350	0351
160	0352	0353	0354	0355	0356	0357	0358	0359	0360	0361	0362	0363	0364	0365	0366	0367
170	0368	0369	0370	0371	0372	0373	0374	0375	0376	0377	0378	0379	0380	0381	0382	0383
180	0384	0385	0386	0387	0388	0389	0390	0391	0392	0393	0394	0395	0396	0397	0398	0399
190	0400	0401	0402	0403	0404	0405	0406	0407	0408	0409	0410	0411	0412	0413	0414	0415
1A0	0416	0417	0418	0419	0420	0421	0422	0423	0424	0425	0426	0427	0428	0429	0430	0431
1B0	0432	0433	0434	0435	0436	0437	0438	0439	0440	0441	0442	0443	0444	0445	0446	0447
1C0	0448	0449	0450	0451	0452	0453	0454	0455	0456	0457	0458	0459	0460	0461	0462	0463
1D0	0464	0465	0466	0467	0468	0469	0470	0471	0472	0473	0474	0475	0476	0477	0478	0479
1E0	0480	0481	0482	0483	0484	0485	0486	0487	0488	0489	0490	0491	0492	0493	0494	0495
1F0	0496	0497	0498	0499	0500	0501	0502	0503	0504	0505	0506	0507	0508	0509	0510	0511
200	0512	0513	0514	0515	0516	0517	0518	0519	0520	0521	0522	0523	0524	0525	0526	0527
210	0528	0529	0530	0531	0532	0533	0534	0535	0536	0537	0538	0539	0540	0541	0542	0543
220	0544	0545	0546	0547	0548	0549	0550	0551	0552	0553	0554	0555	0556	0557	0558	0559
230	0560	0561	0562	0563	0564	0565	0566	0567	0568	0569	0570	0571	0572	0573	0574	0575
240	0576	0577	0578	0579	0580	0581	0582	0583	0584	0585	0586	0587	0588	0589	0590	0591
250	0592	0593	0594	0595	0596	0597	0598	0599	0600	0601	0602	0603	0604	0605	0606	0607
260	0608	0609	0610	0611	0612	0613	0614	0615	0616	0617	0618	0619	0620	0621	0622	0623
270	0624	0625	0626	0627	0628	0629	0630	0631	0632	0633	0634	0635	0636	0637	0638	0639

HEXADECIMAL-DECIMAL INTEGER CONVERSION (continued)

	0	1	2	3	4	5	6	7	8	9	A	B	C	D	E	F
280	0640	0641	0642	0643	0644	0645	0646	0647	0648	0649	0650	0651	0652	0653	0654	0655
290	0656	0657	0658	0659	0660	0661	0662	0663	0664	0665	0666	0667	0668	0669	0670	0671
2A0	0672	0673	0674	0675	0676	0677	0678	0679	0680	0681	0682	0683	0684	0685	0686	0687
2B0	0688	0689	0690	0691	0692	0693	0694	0695	0696	0697	0698	0699	0700	0701	0702	0703
2C0	0704	0705	0706	0707	0708	0709	0710	0711	0712	0713	0714	0715	0716	0717	0718	0719
2D0	0720	0721	0722	0723	0724	0725	0726	0727	0728	0729	0730	0731	0732	0733	0734	0735
2E0	0736	0737	0738	0739	0740	0741	0742	0743	0744	0745	0746	0747	0748	0749	0750	0751
2F0	0752	0753	0754	0755	0756	0757	0758	0759	0760	0761	0762	0763	0764	0765	0766	0767
300	0768	0769	0770	0771	0772	0773	0774	0775	0776	0777	0778	0779	0780	0781	0782	0783
310	0784	0785	0786	0787	0788	0789	0790	0791	0792	0793	0794	0795	0796	0797	0798	0799
320	0800	0801	0802	0803	0804	0805	0806	0807	0808	0809	0810	0811	0812	0813	0814	0815
330	0816	0817	0818	0819	0820	0821	0822	0823	0824	0825	0826	0827	0828	0829	0830	0831
340	0832	0833	0834	0835	0836	0837	0838	0839	0840	0841	0842	0843	0844	0845	0846	0847
350	0848	0849	0850	0851	0852	0853	0854	0855	0856	0857	0858	0859	0860	0861	0862	0863
360	0864	0865	0866	0867	0868	0869	0870	0871	0872	0873	0874	0875	0876	0877	0878	0879
370	0880	0881	0882	0883	0884	0885	0886	0887	0888	0889	0890	0891	0892	0893	0894	0895
380	0896	0897	0898	0899	0900	0901	0902	0903	0904	0905	0906	0907	0908	0909	0910	0911
390	0932	0933	0934	0935	0916	0917	0918	0919	0920	0921	0922	0923	0924	0925	0926	0927
3A0	0928	0929	0930	0931	0912	0913	0914	0915	0926	0937	0938	0939	0940	0941	0942	0943
3B0	0944	0945	0946	0947	0948	0949	0950	0951	0952	0953	0954	0955	0956	0957	0958	0959
3C0	0960	0961	0962	0963	0964	0965	0966	0967	0968	0969	0970	0971	0972	0973	0974	0975
3D0	0976	0977	0978	0979	0980	0981	0982	0983	0984	0985	0986	0987	0988	0989	0990	0991
3E0	0992	0993	0994	0995	0996	0997	0998	0999	1000	1001	1002	1003	1004	1005	1006	1007

400	1024	1025	1026	1027	1028	1029	1030	1031	1032	1033	1034	1035	1036	1037	1038	1039
410	1040	1041	1042	1043	1044	1045	1046	1047	1048	1049	1050	1051	1052	1053	1054	1055
420	1056	1057	1058	1059	1060	1061	1062	1063	1064	1065	1066	1067	1068	1069	1070	1071
430	1072	1073	1074	1075	1076	1077	1078	1079	1080	1081	1082	1083	1084	1085	1086	1087
440	1088	1089	1090	1091	1092	1093	1094	1095	1096	1097	1098	1099	1100	1101	1102	1103
450	1104	1105	1106	1107	1108	1109	1110	1111	1112	1113	1114	1115	1116	1117	1118	1119
460	1120	1121	1122	1123	1124	1125	1126	1127	1128	1129	1130	1131	1132	1133	1134	1135
470	1136	1137	1138	1139	1140	1141	1142	1143	1144	1145	1146	1147	1148	1149	1150	1151
480	1152	1153	1154	1155	1156	1157	1158	1159	1160	1161	1162	1163	1164	1165	1166	1167
490	1168	1169	1170	1171	1172	1173	1174	1175	1176	1177	1178	1179	1180	1181	1182	1183
4A0	1184	1185	1186	1187	1188	1189	1190	1191	1192	1193	1194	1195	1196	1197	1198	1199
4B0	1200	1201	1202	1203	1204	1205	1206	1207	1208	1209	1210	1211	1212	1213	1214	1215
4C0	1216	1217	1218	1219	1220	1221	1222	1223	1224	1225	1226	1227	1228	1229	1230	1231
4D0	1232	1233	1234	1235	1236	1237	1238	1239	1240	1241	1242	1243	1244	1245	1246	1247
4E0	1243	1249	1250	1251	1252	1253	1254	1255	1256	1257	1258	1259	1260	1261	1262	1263
4F0	1264	1265	1266	1267	1268	1269	1270	1271	1272	1273	1274	1275	1276	1277	1278	1279
500	1280	1281	1282	1283	1284	1285	1286	1287	1288	1289	1290	1291	1292	1293	1294	1295
510	1296	1297	1298	1299	1300	1301	1302	1303	1304	1305	1306	1307	1308	1309	1310	1311
520	1312	1313	1314	1315	1316	1317	1318	1319	1320	1321	1322	1323	1324	1325	1326	1327
530	1328	1329	1330	1331	1332	1333	1334	1335	1336	1337	1338	1339	1340	1341	1342	1343
540	1344	1345	1346	1347	1348	1349	1350	1351	1352	1353	1354	1355	1356	1357	1358	1359
550	1360	1361	1362	1363	1364	1365	1366	1367	1368	1369	1370	1371	1372	1373	1374	1375
560	1376	1377	1378	1379	1380	1381	1382	1383	1384	1385	1386	1387	1388	1389	1390	1391
570	1392	1393	1394	1395	1396	1397	1398	1399	1400	1401	1402	1402	1404	1405	1406	1407

HEXADECIMAL-DECIMAL INTEGER CONVERSION (continued)

	0	1	2	3	4	5	6	7	8	9	A	B	C	D	E	F
580	1408	1409	1410	1411	1412	1413	1414	1415	1416	1417	1418	1419	1420	1421	1422	1423
590	1424	1425	1426	1427	1428	1429	1430	1431	1432	1433	1434	1435	1436	1437	1438	1439
5A0	1440	1441	1442	1443	1444	1445	1446	1447	1448	1449	1450	1451	1452	1453	1454	1455
5B0	1456	1457	1458	1459	1460	1461	1462	1463	1464	1465	1466	1467	1468	1469	1470	1471
5C0	1472	1473	1474	1475	1476	1477	1478	1479	1480	1481	1482	1483	1484	1485	1486	1487
5D0	1488	1489	1490	1491	1492	1493	1494	1495	1496	1497	1498	1499	1500	1501	1502	1503
5E0	1504	1505	1506	1507	1508	1509	1510	1511	1512	1513	1514	1515	1516	1517	1518	1519
5F0	1520	1521	1522	1523	1524	1525	1526	1527	1528	1529	1530	1531	1532	1533	1534	1535
600	1536	1537	1538	1539	1540	1541	1542	1543	1544	1545	1546	1547	1548	1549	1550	1551
610	1552	1553	1554	1555	1556	1557	1558	1559	1560	1561	1562	1563	1564	1565	1566	1567
620	1568	1569	1570	1571	1572	1573	1574	1575	1576	1577	1578	1579	1580	1581	1582	1583
630	1584	1585	1586	1587	1588	1589	1590	1591	1592	1593	1594	1595	1596	1597	1593	1599
640	1600	1601	1602	1603	1604	1605	1606	1607	1608	1609	1610	1611	1612	1613	1614	1615
650	1616	1617	1618	1619	1620	1621	1622	1623	1624	1625	1626	1627	1628	1629	1630	1631
660	1632	1633	1634	1635	1636	1637	1638	1639	1640	1641	1642	1643	1644	1645	1646	1647
670	1648	1649	1650	1651	1652	1653	1654	1655	1656	1657	1658	1659	1660	1661	1662	1663
680	1664	1665	1666	1667	1668	1669	1670	1671	1672	1673	1674	1675	1676	1677	1678	1679
690	1680	1681	1682	1683	1684	1685	1686	1687	1688	1682	1690	1691	1692	1693	1694	1695
6A0	1696	1697	1698	1699	1700	1701	1702	1703	1704	1705	1706	1707	1708	1709	1710	1711
6B0	1712	1713	1714	1715	1716	1717	1718	1719	1720	1721	1722	1723	1724	1725	1726	1727
6C0	1728	1729	1730	1731	1732	1733	1734	1735	1736	1737	1738	1739	1740	1741	1742	1743
6D0	1744	1745	1746	1747	1748	1749	1750	1751	1752	1753	1754	1755	1756	1757	1758	1759
6E0	1760	1761	1762	1763	1764	1765	1766	1767	1768	1769	1770	1771	1772	1773	1774	1775
6F0	1776	1777	1778	1779	1780	1781	1782	1783	1784	1785	1786	1787	1788	1789	1790	1791

700	1792	1793	1794	1795	1796	1797	1798	1799	1800	1801	1802	1803	1804	1805	1806	1807
710	1808	1809	1810	1811	1812	1813	1814	1815	1816	1817	1818	1819	1820	1821	1822	1823
720	1824	1825	1826	1827	1828	1829	1830	1831	1832	1833	1834	1835	1836	1837	1838	1839
730	1840	1841	1842	1843	1844	1845	1846	1847	1848	1849	1850	1851	1852	1853	1854	1855
740	1856	1857	1858	1859	1860	1861	1862	1863	1864	1865	1866	1867	1868	1869	1870	1871
750	1872	1873	1874	1875	1876	1877	1878	1879	1880	1881	1882	1883	1884	1885	1886	1887
760	1888	1889	1890	1891	1892	1893	1894	1895	1896	1897	1898	1899	1900	1901	1902	1903
770	1904	1905	1906	1907	1908	1909	1910	1911	1912	1913	1914	1915	1916	1917	1918	1919
780	1920	1921	1922	1923	1924	1925	1926	1927	1928	1929	1930	1931	1932	1933	1934	1935
790	1936	1937	1938	1939	1940	1941	1942	1943	1944	1945	1946	1947	1948	1949	1950	1951
7A0	1952	1953	1954	1955	1956	1957	1958	1959	1960	1961	1962	1963	1964	1965	1966	1967
7B0	1968	1969	1970	1971	1972	1973	1974	1975	1976	1977	1978	1979	1980	1981	1982	1983
7C0	1984	1985	1986	1987	1988	1989	1990	1991	1992	1993	1994	1995	1996	1997	1998	1999
7D0	2000	2001	2002	2003	2004	2005	2006	2007	2008	2009	2010	2011	2012	2013	2014	2015
7E0	2016	2017	2018	2019	2020	2021	2022	2023	2024	2025	2026	2027	2028	2029	2030	2031
7F0	2032	2033	2034	2035	2036	2037	2038	2039	2040	2041	2042	2043	2044	2045	2046	2047
800	2048	2049	2050	2051	2052	2053	2054	2055	2056	2057	2058	2059	2060	2061	2062	2063
810	2064	2065	2066	2067	2068	2069	2070	2071	2072	2073	2074	2075	2076	2077	2078	2079
820	2080	2081	2082	2083	2084	2085	2086	2087	2088	2089	2090	2091	2092	2093	2094	2095
830	2096	2097	2098	2099	2100	2101	2102	2103	2104	2105	2106	2107	2108	2109	2110	2111
840	2112	2113	2114	2115	2116	2117	2118	2119	2120	2121	2122	2123	2124	2125	2126	2127
850	2128	2129	2130	2131	2132	2133	2134	2135	2136	2137	2138	2139	2140	2141	2142	2143
860	2144	2145	2146	2147	2148	2149	2150	2151	2152	2153	2154	2155	2156	2157	2158	2159
870	2160	2161	2162	2163	2164	2165	2166	2167	2168	2169	2170	2171	2172	2173	2174	2175

HEXADECIMAL-DECIMAL INTEGER CONVERSION (continued)

	0	1	2	3	4	5	6	7	8	9	A	B	C	D	E	F
880	2176	2177	2178	2179	2180	2181	2182	2183	2184	2185	2186	2187	2188	2189	2190	2191
890	2192	2193	2194	2195	2196	2197	2198	2199	2200	2201	2202	2203	2204	2205	2206	2207
8A0	2208	2209	2210	2211	2212	2213	2214	2215	2216	2217	2218	2219	2220	2221	2222	2223
8B0	2224	2225	2226	2227	2228	2229	2230	2231	2232	2233	2234	2235	2236	2237	2238	2239
8C0	2240	2241	2242	2243	2244	2245	2246	2247	2248	2249	2250	2251	2252	2253	2254	2255
8D0	2256	2257	2258	2259	2260	2261	2262	2263	2264	2265	2266	2267	2268	2269	2270	2271
8E0	2272	2273	2274	2275	2276	2277	2278	2279	2280	2281	2282	2283	2284	2285	2286	2287
8F0	2288	2289	2290	2291	2292	2293	2294	2295	2296	2297	2298	2299	2300	2301	2302	2303
900	2304	2305	2306	2307	2308	2309	2310	2311	2312	2313	2314	2315	2316	2317	2318	2319
910	2320	2321	2322	2323	2324	2325	2326	2327	2328	2329	2330	2331	2332	2333	2334	2335
920	2336	2337	2338	2339	2340	2341	2342	2343	2344	2345	2346	2347	2348	2349	2350	2351
930	2352	2353	2354	2355	2356	2357	2358	2359	2360	2361	2362	2363	2364	2365	2366	2367
940	2368	2369	2370	2371	2372	2373	2374	2375	2376	2377	2378	2379	2380	2381	2382	2383
950	2384	2385	2386	2387	2383	2389	2390	2391	2392	2393	2394	2395	2396	2397	2398	2399
960	2400	2401	2402	2403	2404	2405	2406	2307	2408	2409	2410	2411	2412	2413	2414	2415
970	2416	2417	2418	2419	2420	2421	2422	2423	2424	2425	2426	2427	2428	2429	2430	2431
980	2432	2433	2434	2435	2436	2437	2438	2439	2440	2441	2442	2443	2444	2445	2446	2447
990	2448	2449	2450	2451	2452	2453	2454	2455	2456	2457	2458	2459	2460	2461	2462	2463
9A0	2464	2465	2466	2467	2468	2469	2470	2471	2472	2473	2474	2475	2476	2477	2478	2479
9B0	2480	2481	2482	2483	2484	2485	2486	2487	2488	2489	2490	2491	2492	2493	2494	2495
9C0	2496	2497	2498	2499	2500	2501	2502	2503	2504	2505	2506	2507	2508	2509	2510	2511
9D0	2512	2513	2514	2515	2516	2517	2518	2519	2520	2521	2522	2523	2524	2525	2526	2527
9E0	2528	2529	2530	2531	2532	2533	2534	2535	2536	2537	2538	2539	2540	2541	2542	2543
9F0	2544	2545	2546	2547	2548	2549	2550	2551	2552	2553	2554	2555	2556	2557	2558	2559

A00	2560	2561	2562	2563	2564	2565	2566	2567	2568	2569	2570	2571	2572	2573	2574	2575
A10	2576	2577	2578	2579	2580	2581	2582	2583	2584	2585	2586	2587	2588	2589	2590	2591
A20	2592	2593	2594	2595	2596	2597	2598	2599	2600	2601	2602	2603	2604	2605	2606	2607
A30	2608	2609	2610	2611	2612	2613	2614	2615	2616	2617	2618	2619	2620	2621	2622	2623
A40	2624	2625	2626	2627	2628	2629	2630	2631	2632	2633	2634	2635	2636	2637	2638	2639
A50	2640	2641	2642	2643	2644	2645	2646	2647	2648	2649	2650	2651	2652	2653	2654	2655
A60	2656	2657	2658	2659	2660	2661	2662	2663	2664	2665	2666	2667	2668	2669	2670	2671
A70	2672	2673	2674	2675	2676	2677	2678	2679	2680	2681	2682	2683	2684	2685	2686	2687
A80	2688	2689	2690	2691	2692	2693	2694	2695	2696	2697	2698	2699	2700	2701	2702	2703
A90	2704	2705	2706	2707	2708	2709	2710	2711	2712	2713	2714	2715	2716	2717	2718	2719
AA0	2720	2721	2722	2723	2724	2725	2726	2727	2728	2729	2730	2731	2732	2733	2734	2735
AB0	2736	2737	2738	2739	2740	2741	2742	2743	2744	2745	2746	2747	2748	2749	2750	2751
AC0	2752	2753	2754	2755	2756	2757	2758	2759	2760	2761	2762	2763	2764	2765	2766	2767
AD0	2768	2769	2770	2771	2772	2773	2774	2775	2776	2777	2778	2779	2780	2781	2782	2783
AE0	2784	2785	2786	2787	2788	2789	2790	2791	2792	2793	2794	2795	2796	2797	2798	2799
AF0	2800	2801	2802	2803	2804	2805	2806	2807	2808	2809	2810	2811	2812	2813	2814	2815
B00	2816	2817	2818	2819	2820	2821	2822	2823	2824	2825	2826	2827	2828	2829	2830	2831
B10	2832	2833	2834	2835	2836	2837	2838	2839	2840	2841	2842	2843	2844	2845	2846	2847
B20	2848	2849	2850	2851	2852	2853	2854	2855	2856	2857	2858	2859	2860	2861	2862	2863
B30	2864	2865	2866	2867	2868	2869	2870	2871	2872	2873	2874	2875	2876	2877	2878	2879
B40	2880	2881	2882	2883	2884	2885	2886	2887	2888	2889	2890	2891	2892	2893	2894	2895
B50	2896	2897	2898	2899	2900	2901	2902	2903	2904	2905	2906	2907	2908	2909	2910	2911
B60	2912	2913	2914	2915	2916	2917	2918	2919	2920	2921	2922	2923	2924	2925	2926	2927
B70	2928	2929	2930	2931	2932	2933	2934	2935	2936	2937	2938	2939	2940	2941	2942	2943

HEXADECIMAL-DECIMAL INTEGER CONVERSION (continued)

	0	1	2	3	4	5	6	7	8	9	A	B	C	D	E	F
B80	2944	2945	2946	2947	2948	2949	2950	2951	2952	2953	2954	2955	2956	2957	2985	2959
B90	2960	2961	2962	2963	2964	2965	2966	2967	2968	2969	2970	2971	2972	2973	2974	2975
BA0	2976	2977	2978	2979	2980	2981	2982	2983	2984	2985	2986	2987	2988	2989	2990	2991
BB0	2992	2993	2994	2995	2996	2997	2998	2999	3000	3001	3002	3003	3004	3005	3006	3007
BC0	3008	3009	3010	3011	3012	3013	3014	3015	3016	3017	3018	3019	3020	3021	3022	3023
BD0	3024	3025	3026	3027	3028	3029	3030	3031	3032	3033	3034	3035	3036	3037	3038	3039
BE0	3040	3041	3042	3043	3044	3045	3046	3047	3048	3049	3050	3051	3052	3053	3054	3055
BF0	3056	3057	3058	3059	3060	3061	3062	3063	3064	3065	3066	3067	3068	3069	3070	3071
C00	3072	3073	3074	3075	3076	3077	3078	3079	3080	3081	3082	3083	3084	3085	3086	3087
C10	3088	3089	3090	3091	3092	3093	3094	3095	3096	3097	3098	3099	3100	3101	3102	3103
C20	3104	3105	3106	3107	3108	3109	3110	3111	3112	3113	3114	3115	3116	3117	3118	3119
C30	3120	3121	3122	3123	3124	3125	3126	3127	3128	3129	3130	3131	3132	3133	3134	3135
C40	3136	3137	3138	3139	3140	3141	3142	3143	3144	3145	3146	3147	3148	3149	3150	3151
C50	3152	3153	3154	3155	3156	3157	3158	3159	3160	3161	3162	3163	3164	3165	3166	3167
C60	3168	3169	3170	3171	3172	3173	3174	3175	3176	3177	3178	3179	3180	3181	3182	3183
C70	3184	3185	3186	3187	3188	3189	3190	3191	3192	3193	3194	3195	3196	3197	3198	3199
C80	3200	3201	3202	3203	3204	3205	3206	3207	3208	3209	3210	3211	3212	3213	3214	3215
C90	3216	3217	3218	3219	3220	3221	3222	3223	3224	3225	3226	3227	3228	3229	3230	3231
CA0	3232	3233	3234	3235	3236	3237	3238	3239	3240	3241	3242	3243	3244	3245	3246	3247
CB0	3248	3249	3250	3251	3252	3253	3254	3255	3256	3257	3258	3259	3260	3261	3262	3263
CC0	3264	3265	3266	3267	3268	3269	3270	3271	3272	3273	3274	3275	3276	3277	3278	3279
CD0	3280	3281	3282	3283	3284	3285	3286	3287	3288	3289	3290	3291	3292	3293	3294	3295
CE0	3296	3297	3298	3299	3300	3301	3302	3303	3304	3305	3306	3307	3308	3309	3310	3311
CF0	3312	3313	3314	3315	3316	3317	3318	3319	3320	3321	3322	3323	3324	3325	3326	3327

D00	3328	3329	3330	3331	3332	3333	3334	3335	3336	3337	3338	3339	3340	3341	3342	3343
D10	3344	3345	3346	3347	3348	3349	3350	3351	3352	3353	3354	3355	3356	3357	3358	3359
D20	3360	3361	3362	3363	3364	3365	3366	3367	3368	3369	3370	3371	3372	3373	3374	3375
D30	3376	3377	3378	3379	3380	3381	3382	3383	3384	3385	3386	3387	3388	3389	3390	3391
D40	3392	3393	3394	3395	3396	3397	3398	3399	3400	3401	3402	3403	3404	3405	3406	3407
D50	3408	3409	3410	3411	3412	3413	3414	3415	3416	3417	3418	3419	3420	3421	3422	3423
D60	3424	3425	3426	3427	3428	3429	3430	3431	3432	3433	3434	3435	3436	3437	3438	3439
D70	3440	3441	3442	3443	3444	3445	3446	3447	3448	3449	3450	3451	3452	3453	3454	3455
D80	3456	3457	3458	3459	3460	3461	3462	3463	3464	3465	3466	3467	3468	3469	3470	3471
D90	3472	3473	3474	3475	3476	3477	3478	3479	3480	3481	3482	3483	3484	3485	3486	3487
DA0	3488	3489	3490	3491	3492	3493	3494	3495	3496	3497	3498	3499	3500	3501	3502	3503
DB0	3504	3505	3506	3507	3508	3509	3510	3511	3512	3513	3514	3515	3516	3517	3518	3519
DC0	3520	3521	3522	3523	3524	3525	3526	3527	3528	3529	3530	3531	3532	3533	3534	3535
DD0	3536	3537	3538	3539	3540	3541	3542	3543	3544	3545	3546	3547	3548	3549	3550	3551
DE0	3552	3553	3554	3555	3556	3557	3558	3559	3560	3561	3562	3563	3564	3565	3566	3567
DF0	3568	3569	3570	3571	3572	3573	3574	3575	3576	3577	3578	3579	3580	3581	3582	3583
E00	3584	3585	3586	3587	3588	3589	3590	3591	3592	3593	3594	3595	3596	3597	3598	3599
E10	3600	3601	3602	3603	3604	3605	3606	3607	3608	3609	3610	3611	3612	3613	3614	3615
E20	3616	3617	3618	3619	3620	3621	3622	3623	3624	3625	3626	3627	3628	3629	3630	3631
E30	3632	3633	3634	3635	3636	3637	3638	3639	3640	3641	3642	3643	3644	3645	3646	3647
E40	3648	3649	3650	3651	3652	3653	3654	3655	3656	3657	3658	3659	3660	3661	3662	3663
E50	3664	3665	3666	3667	3668	3669	3670	3671	3672	3673	3674	3675	3676	3677	3678	3679
E60	3680	3681	3682	3683	3684	3685	3686	3687	3688	3689	3690	3691	3692	3693	3694	3695
E70	3696	3697	3698	3699	3700	3701	3702	3703	3704	3705	3706	3707	3708	3709	3710	3711

HEXADECIMAL-DECIMAL INTEGER CONVERSION (continued)

	0	1	2	3	4	5	6	7	8	9	A	B	C	D	E	F
E80	3712	3713	3714	1715	3716	3717	3718	3719	3720	3721	3722	3723	3724	3725	3726	3727
E90	3728	3729	3730	3731	3732	3733	3734	3735	3736	3737	3738	3739	3740	3741	3742	3743
EA0	3744	3745	3746	3747	3748	3749	3750	3751	3752	3753	3754	3755	3756	3757	3758	3759
EB0	3760	3761	3762	3763	3764	3765	3766	3767	3768	3769	3770	3771	3772	3773	3774	3775
EC0	3776	3777	3778	3779	3780	3781	3782	3783	3784	3785	3786	3787	3788	3789	3790	3791
ED0	3792	3793	3794	3795	3796	3797	3798	3799	3800	3801	3802	3803	3804	3805	3806	3807
EE0	3808	3809	3810	3811	3812	3813	3814	3815	3816	3817	3818	3819	3820	3821	3822	3823
EF0	3824	3825	3826	3827	3828	3829	3830	3831	3832	3833	3834	3835	3836	3837	3838	3839
F00	3840	3841	3842	3843	3844	3845	3846	3847	3848	3849	3850	3851	3852	3853	3854	3855
F10	3856	3857	3858	3859	3860	3861	3862	3863	3864	3865	3866	3867	3868	3869	3870	3871
F20	3872	3873	3874	3875	3876	3877	3878	3879	3880	3881	3882	3563	3884	3885	3886	3887
F30	3888	3889	3890	3891	3892	3893	3894	3895	3896	3897	3898	3899	3900	3901	3902	3903
F40	3904	3905	3906	3907	3908	3909	3910	3911	3912	2313	3914	3915	3916	3917	3918	3919
F50	3920	3921	3922	3923	3924	3925	3926	3927	3928	3929	3930	3931	3932	3933	3934	3935
F60	3936	3937	3938	3939	3940	3941	3942	3943	3944	3945	394	3947	3948	3949	3950	3951
F70	3952	3953	3954	3955	3956	3957	3958	3959	3960	3961	3962	3963	3964	3965	3966	3967
F80	3968	3969	3970	3971	3972	3973	3974	3975	3976	3977	3978	3979	3980	3981	3982	3983
F90	3984	3985	3986	3987	3988	3989	3990	3991	3992	3993	3994	3995	3996	3997	3998	3999
FA0	4000	4001	4002	4003	4004	4005	4006	4007	4008	4009	4010	4011	4012	4013	4014	4015
FB0	4016	1017	1018	4019	4020	4021	4022	4023	4024	4025	4026	4027	4028	4029	4030	4031
FC0	4032	4033	4034	4035	4036	4037	4038	4039	4040	4041	4042	4043	4044	4045	4046	4047
FD0	4048	4049	4050	4051	4052	4053	4054	4055	4056	4057	4058	4059	4060	4061	4062	4063
FE0	4064	4065	4066	4067	4068	4069	4070	4071	4072	4073	4074	4075	4076	4077	4078	4079
FF0	4080	4081	4082	4083	4084	4085	4086	4087	4088	4089	4090	4091	4092	4093	4094	4095

Index

Index

A

Abounds 66
Accumulator 171
Address, last 172
Address latch enable 72
Address, lower 72
    pseudo-16-bit 73
    upper 72
Address zero 186
Addressing 72
Alignment procedure 104
Analog 110
Audigen 11
Auxiliary input 164

B

Ball-bearings 13
Basic programmer, construction 144
Battery monitor 130, 190
Battery monitor routine 191
Brain board 13
Brain board assembly 81

C

Cease motion 116
Centralization 18, 65
Charging system 62
Circuit assembly 126
Circuit construction 116
Circuit description 101, 122
Circuit testing 116, 128
Circuit wiring 88
Command/status register 73
Comparison 110
Connectors 43
Contact switches 135
Control-oriented 68
Crosstalk 60
Current requirements 50

D

Data pitches 110
Dead axle 33
Design considerations 121
Digital record 85
Display stage 164
Dual-mode 80

E

Exactitude 110
Excom 11

F

Feeding-time 58
Final programming notes 201
First reference pitch 192
Fredian 109
Fredian grammar 109
Frequency shift keying 149

G

Gearmotor, drive 29
    steering 29

H

Hexadecimal 145

I

Identification 110
Information-oriented 68
Initialization 186
Instruction set 173
Instruction set, using 175
Interface 36
Interface retrieval 159
Interrupt 78
Interrupt addresses 190
Interrupt stack 183
Interrupt system 172

K

Kansas City Standard 148
Kilobytes 68

L

Logic stepper motors 26

M

Mainframe 16, 24
Manual programmer 186, 201
Manual store 201
MCS —5/8 User's Manual 81
Memory 66
Memory, display 68
    random access 147
Modems 148
Modularity 43
Motorframe 16,24
Motors, mounting 29

N

Name code 116

O

Open-collector 82
Oscillation 99

P

Passes 85
Playback mode 166
Point of reference 58
Port allocation 74
Ports, I/O 67
Power system 12
Power system, checking 56
Probability number 196
Probability table 196
Processor 67
Processor and instructions 170
Processor interaction 66
Program counter 73, 172
Programming 21
Programmability 66
Programmer, deluxe 144
    final testing 145
Programming tool, tape interface unit 21
Pulse width modulation 149

R

Random mode 20
Random number generator 196
Rating, ampere-hour 50
RCU-85 68
RCU-85 inputs, others 78
Reduced overshoot 38
Reference frequency 110, 119
Regulator construction 53
Relays, direction 38
    mounting 39
    selection 36
    wiring 38
Remote control 108
Reset switches 138
Resolution 110
Retrieval function 158
Retrieve mode 156
Robot controller unit 69
Robot, choice of goals 20
    choice of motors 25
    definition 9

design considerations....23
goal-oriented design....16
final test....34
optimum....201
ultimate....201
using this voulme....12
Robot pet in every home....214
Routine, Arasem....196
Callable....197
Excom....192
watchdog....200

S

Saturation level....52
Sensorframe....39
Series pass element....52
Servo potentiometer....40
Servodrive....41
Servodrive description....85
Software....168
Software, autocritical....214
fear of....169
Soniscan....10, 93
Soniscan design....94
Soniscan in operation....106
Speech recognition....109
Stack pointer....172
Status display device....118
Storage function....152
Subroutines....183
Subroutine, Candidate....196
general....200
Sync flag....114
Sync space....192

T

Tape interface....146, 147, 201
Tape interface, circuit description....149
construction....164
design considerations....147
operating....152
testing....164
Timers....77

Timers, event 78, 133, 191
Toward better locomotion 209
Toward enhanced software 212
Toward improved memory 210
Transducers 97
Transmit pulse 102
Tristate buffer 153

U

Ultimate improvement 47
Ultrasonic 94

V

Validity period 194
Various fine touches 136
Versatility 65
Voltage regulators 49, 52
Voltage requirements 49

W

Wheels 32

Z

Zero crossings 114